Speed Math

by Gaurav Tekriwal

ALPHA
A member of Penguin Group (USA) Inc.

ALPHA BOOKS

Published by Penguin Group (USA) Inc.

Penguin Group (USA) Inc., 375 Hudson Street, New York, New York 10014, USA • Penguin Group (Canada), 90 Eglinton Avenue East, Suite 700, Toronto, Ontario M4P 2Y3, Canada (a division of Pearson Penguin Canada Inc.) • Penguin Books Ltd., 80 Strand, London WC2R 0RL, England • Penguin Ireland, 25 St. Stephen's Green, Dublin 2, Ireland (a division of Penguin Books Ltd.) • Penguin Group (Australia), 250 Camberwell Road, Camberwell, Victoria 3124, Australia (a division of Pearson Australia Group Pty. Ltd.) • Penguin Books India Pvt. Ltd., 11 Community Centre, Panchsheel Park, New Delhi—110 017, India • Penguin Group (NZ), 67 Apollo Drive, Rosedale, North Shore, Auckland 1311, New Zealand (a division of Pearson New Zealand Ltd.) • Penguin Books (South Africa) (Pty.) Ltd., 24 Sturdee Avenue, Rosebank, Johannesburg 2196, South Africa • Penguin Books Ltd., Registered Offices: 80 Strand, London WC2R 0RL, England

International Standard Book Number: 978-1-61564-316-5
Library of Congress Catalog Card Number: 2013945257

18 17 16 8 7 6 5 4 3

Interpretation of the printing code: The rightmost number of the first series of numbers is the year of the book's printing; the rightmost number of the second series of numbers is the number of the book's printing. For example, a printing code of 14-1 shows that the first printing occurred in 2014.

Printed in the United States of America

Publisher: *Mike Sanders*
Executive Managing Editor: *Billy Fields*
Senior Acquisitions Editor: *Tom Stevens*
Development Editor: *Kayla Dugger*
Senior Production Editor/Proofreader: *Janette Lynn*
Cover/Book Designer: *William Thomas*
Indexer: *Brad Herriman*
Layout: *Brian Massey*

Contents

Appendixes

Introduction

Can you multiply 98 × 97 in your head in less than five seconds?

You may think doing such a problem in your head is a pipe dream, but it's not. In fact, doing math without a calculator is critical. In places such as the United Kingdom, calculators are being banned from math tests. So if you can't do the math in your head or on a piece of paper, you're stuck.

This is where speed math comes in. By using the methods in this book, you'll become more adept at solving problems, including the following:

- Addition and subtraction
- Multiplication and division
- Fractions and decimals
- Squares and square roots
- Numbers to different powers

And that doesn't cover everything you'll learn. In fact, by the end of this book, you'll be amazed at how quick and easy it is to solve almost any problem.

I hope you enjoy your journey into speed math.

How This Book Is Organized

This book is divided into two parts. The chapters in each part don't have to be read in order, so if you have a type of math that's of particular interest to you, feel free to skip to it.

Part 1, High-Speed Operations, provides easier ways of doing multiplication, addition, subtraction, and division. I lead off with multiplication, as it tends to have the most exciting tricks. I end this part by showing you how to check your answers using digit sums.

Part 2, High-Speed Math Applications, tackles decimals, fractions, and percentages. I also discuss how to find the value of powers and roots.

The appendixes include practice problems and answers to help you further grasp the concepts from each chapter.

Extras

Look for these handy sidebars throughout the book. They have important information you need to guide you through doing speed math.

These sidebars give you shortcuts and advice on different speed math methods.

Whoa! Watch out for these sidebars, which tell you about potential pitfalls you may encounter while doing your calculations.

But Wait! There's More!

Have you logged on to idiotsguides.com lately? If you haven't, go there now! As a bonus to the book, we've included tons of additional sample problems and other speed-math tricks you'll want to check out, all online. Point your browser to idiotsguides.com/speedmath and enjoy!

Acknowledgments

I would like to first express my gratitude to my guru and teacher, His Holiness Jagadguru Swami Shri Nischalananda Saraswati, the current Shankaracharya of Puri. He is the current global ambassador of Vedic Mathematics, and I owe him deeply for his valuable guidance and priceless teachings.

I would like to thank my parents, who have supported me and had faith in me through the last 13 years of my journey with Vedic Mathematics. I would also like to thank my wife, Shree, and my close friend, Varun Poddar, who have been pillars of support and encouragement, especially at times when I most needed it.

I would like to thank Carole Jelen of Waterside Productions and Alpha Books for considering me worthy to write this book. I would also like to thank my editors, Tom Stevens and Kayla Dugger, for their patience and support along the way.

Thanks to Manas Das and Amitava Ghosh in India, who helped me with the figure designs presented in this book.

Last but not least, I beg forgiveness of all those who have been with me over the course of the years and whose names I have failed to mention. Thank you.

Trademarks

PART

1

High-Speed Operations

You know your basic operations—addition, subtraction, multiplication, and division. But I'm sure you've wondered many times if there were quicker and easier ways to get the answers. Well, you're in luck! Through speed math, you can use shortcuts and other helpful methods to not only get the fast answers, but, more importantly, the *right* answers!

In this part, I guide you through the ways you can speed up addition, subtraction, multiplication, and division. I close the part with a chapter on digit sums, which can help you check your answers for each operation.

CHAPTER

1

Multiplication

In This Chapter

- Getting "hands-on" with finger multiplication
- Tricks you can use when multiplying by 11
- How the base method can help you simplify multiplication problems
- Multiplying numbers that aren't close to a base
- Using the vertically and crosswise method to multiply numbers

You may wonder why I'm beginning with multiplication. After all, wouldn't it be easier to start with addition or subtraction? It's that sense of dread you may feel about jumping into multiplication that makes me want to talk to you about it right away. Speed math is especially impressive for multiplication problems; the methods are so simple, you'll actually enjoy doing it!

In this chapter, I talk about the many ways you can quickly and easily solve multiplication problems, including finger multiplication, a special technique when multiplying numbers by 11, the base method, and the vertically and crosswise method.

Finger Multiplication

One of the most daunting problems for students—and adults—is getting their times table correct. In different countries, math teachers rely on different methods to make this daunting task easier. The most popular method is rote memorization. Some also rely on flash cards or visual representation of the figures.

I want to show you a way to do multiplication tables that only requires your hands—it's called *finger multiplication*. In my experience, people find finger multiplication very enjoyable, as it's very tactile and gives instant answers.

But how does finger multiplication work? If you open up your palms and look at your fingers closely, you'll note that each finger (excluding the thumb) can be divided into four parts.

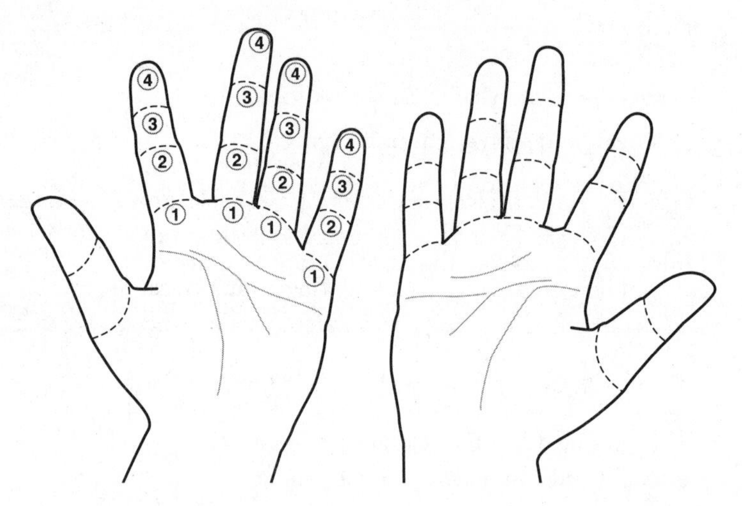

You can use these finger divides or finger joints on both your hands to do basic calculations for your multiplication tables.

Note that on one hand you can actually count up to 16, as you can see in the following image.

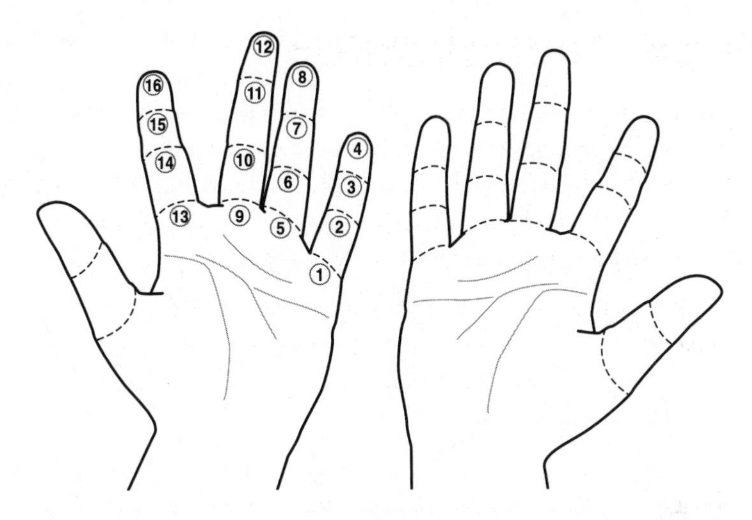

Now that you know how the finger joints play a role, let me take you through how to do the multiplication with different times tables.

The 2 Times Table

To do finger multiplication for the 2 times table, group the finger joints in twos and skip count by those to get your answers.

Say you have to solve 2 × 1. Using finger multiplication, all you need to do is count in twos once. This gives you the answer of 2.

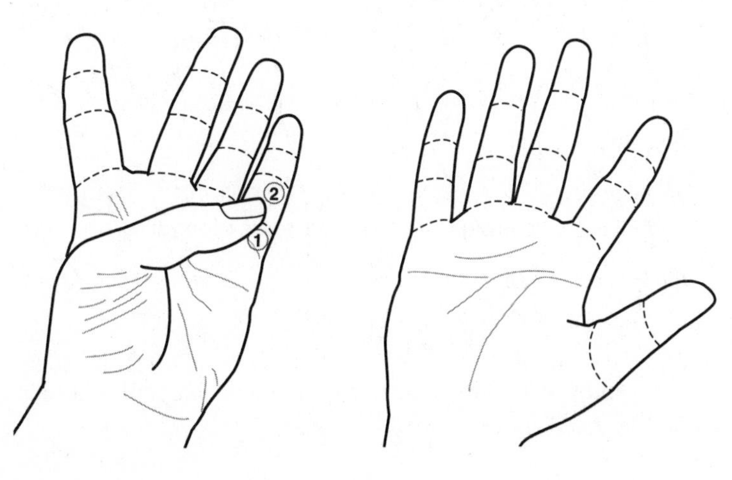

For 2 × 5, you count in twos five times to get 10. Because you know there are two groups 2 on each hand, you can skip count in twos and reach 10.

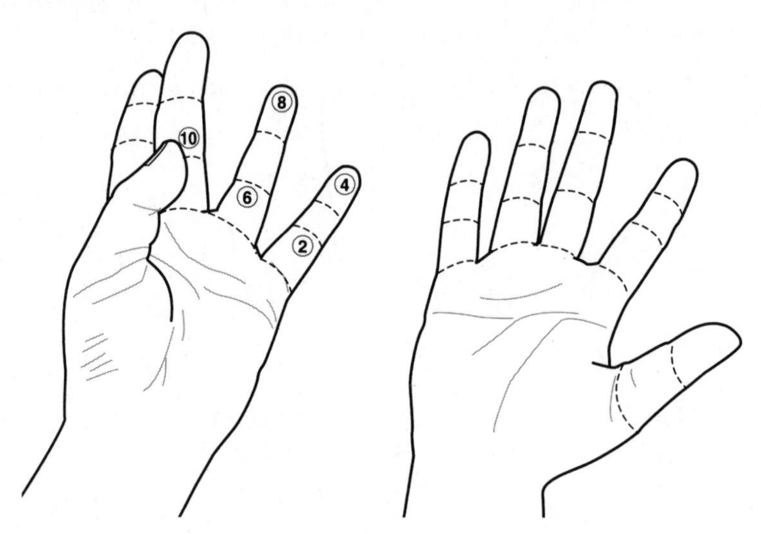

And 2 × 8 is equal to 16.

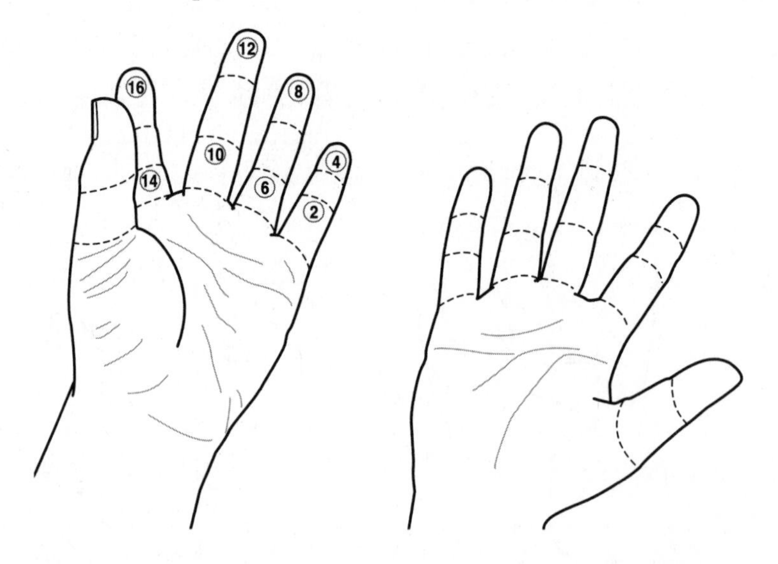

You can use the same process on the fingers on your other hand to do 2 × 9 and above.

The 3 Times Table

What you did for the 2 times table can be extended to the 3 times table as well. You use the joints to make groups of three and skip count by those to arrive at your answer.

For example, to get 3 × 1, you count by threes once, which tells you the answer is 3.

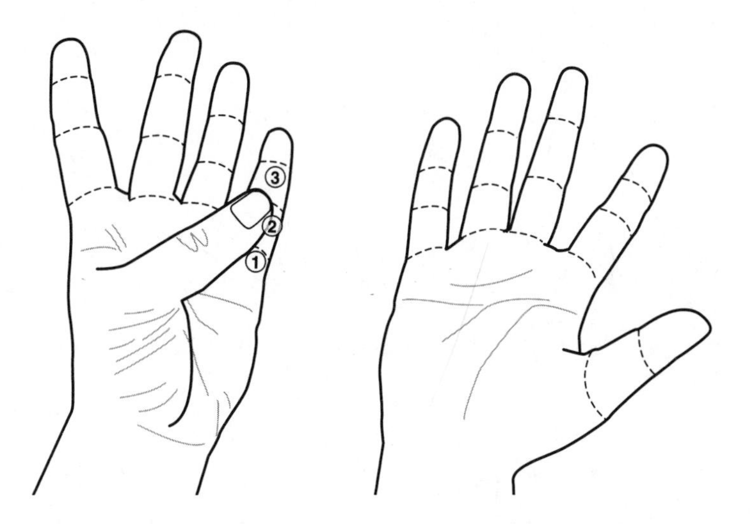

3 × 2 means you count in threes twice. This gives you an answer of 6.

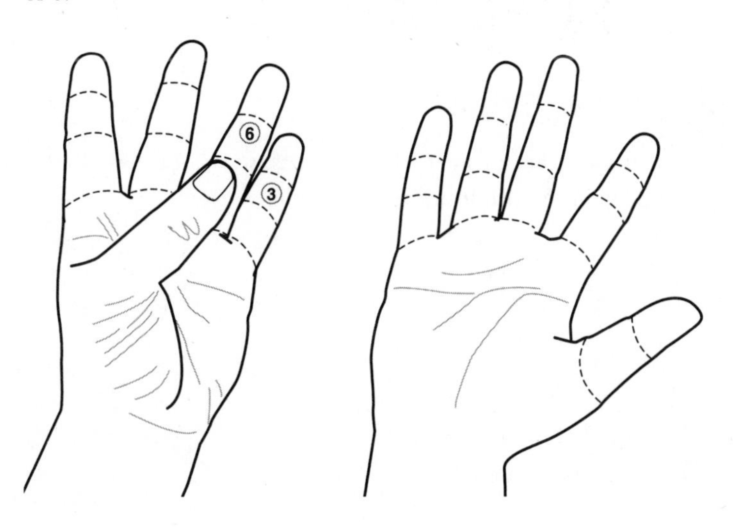

Similarly, for 3 × 3, you count in threes three times to get your answer of 9.

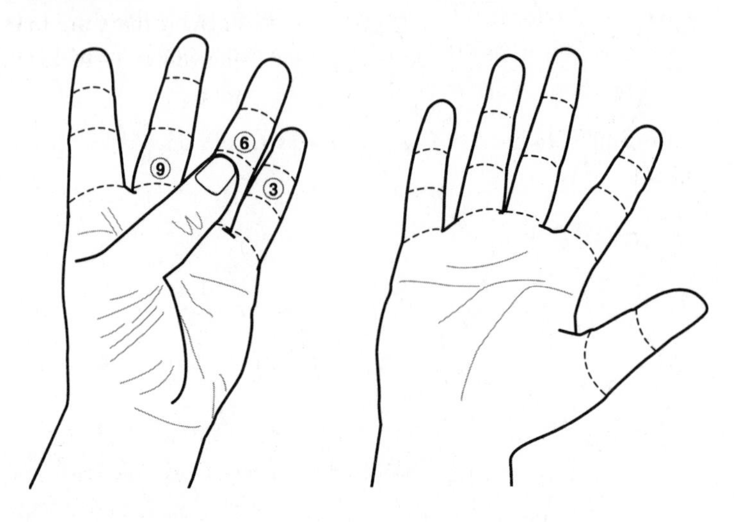

For 3 × 6 and upward, you can use your other hand for your calculations.

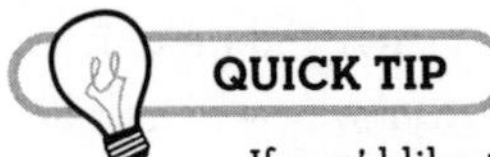

If you'd like to have the 3 times table spread more evenly on your hands, consider counting only the first three joints on each finger and thumb.

The 4 Times Table

As you know, each finger has four parts, so the 4 times table is the easiest finger multiplication you can do.

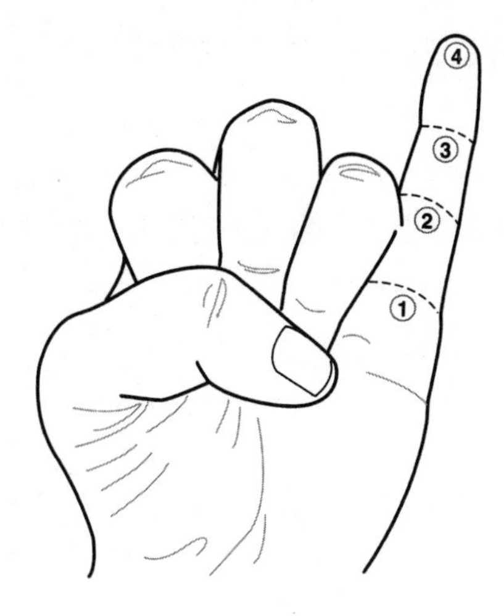

To do 4 × 1, you simply hold out the little finger. Because there are four on one finger, you have your answer: 4.

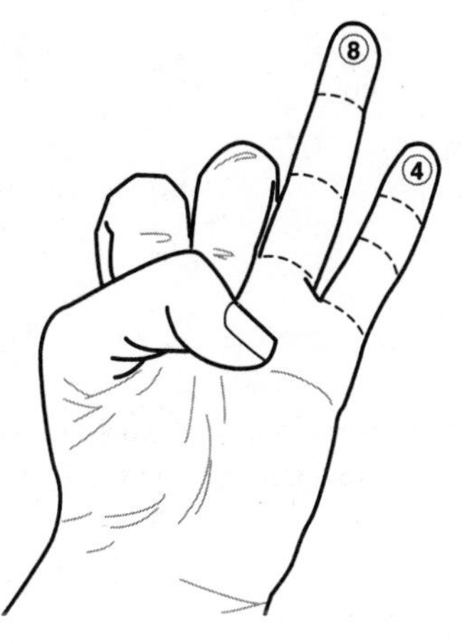

Similarly, for 4 × 2, you hold out two fingers. You can count in fours twice on each finger, or you can just simply think of how there are four joints on each and quickly get your answer, which is 8.

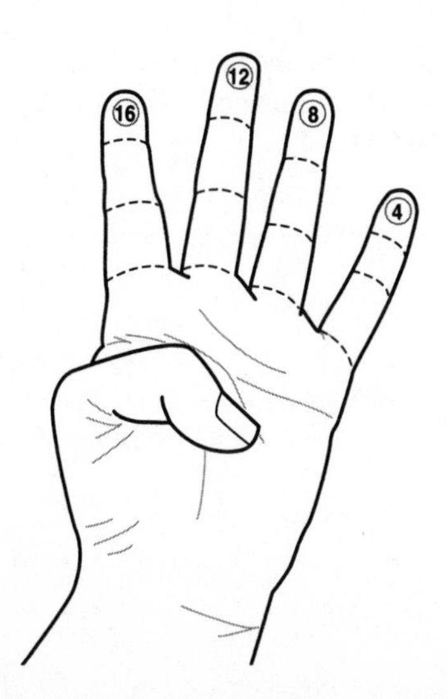

4 × 4 requires you lift four fingers to get 16.

The Universal Times Tables

To understand the universal times tables in finger multiplication, you already need to know your times tables up to 5. This method can help you do 6 times, 7 times, 8 times, 9 times, and 10 times tables.

To do the multiplication for these tables, you mentally number the fingers of both your hands from 10 to 6:

Thumb: 10

Index finger: 9

Middle finger: 8

Ring finger: 7

Little finger: 6

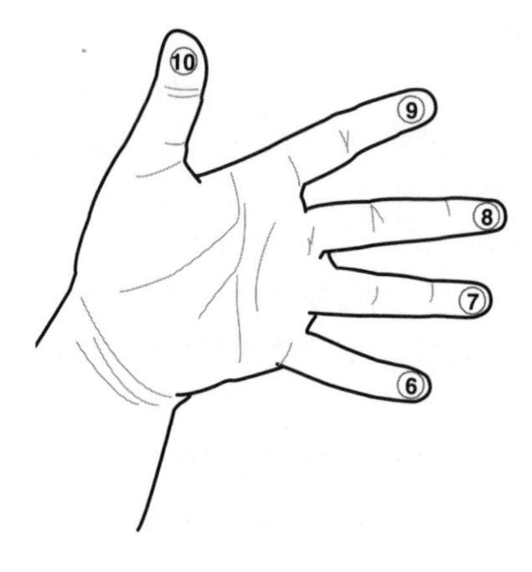

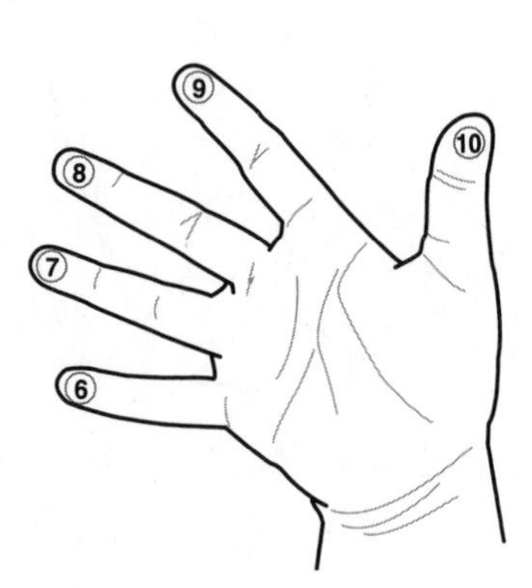

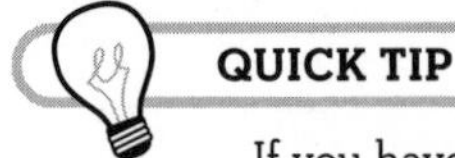

If you have trouble envisioning the numbers on your own fingers, feel free to use a pen to label them. Just make sure you don't wear them on your hands when taking a test!

Now let's try a problem, 9 × 8. Join the 9 on your left hand to the 8 on your right hand as shown.

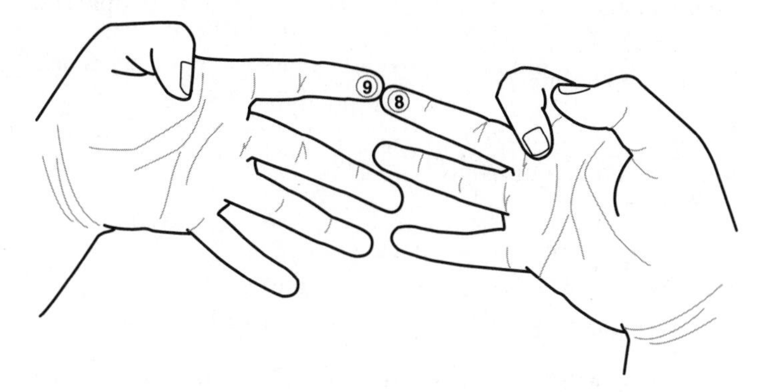

You count the joined fingers and the fingers below the joined fingers to get the first part of your answer. Here, you have a total of seven fingers, so the first digit is 7.

You use the fingers above your joined fingers to find the last digit. On top, you have one finger on the left side and two fingers on the right side. To get the last digit of your answer, you multiply 1 × 2 and get 2.

You then combine 7 and 2, which gives you 72 as your answer.

Let me give you another example to try, 6 × 9. First, join the 6 on your left hand with the 9 on your right hand.

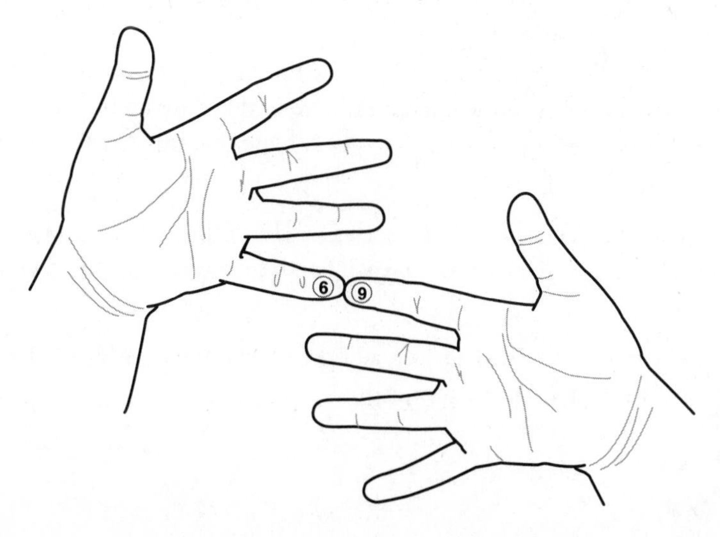

Again, you count the joined fingers and the finger below to get the first part of your answer. Here, it's 5.

You then count the fingers above the joined fingers on each hand separately and multiply. In this case, you have four fingers on one and one on the other, which multiplied together gives you 4.

After joining the two parts of the answer together, your complete answer becomes 54.

One thing that may come up while doing your universal times tables is carryovers. Let me give you an example to show you how to deal with them.

Say you want to solve 6 × 6. For this problem, both your little fingers are joined.

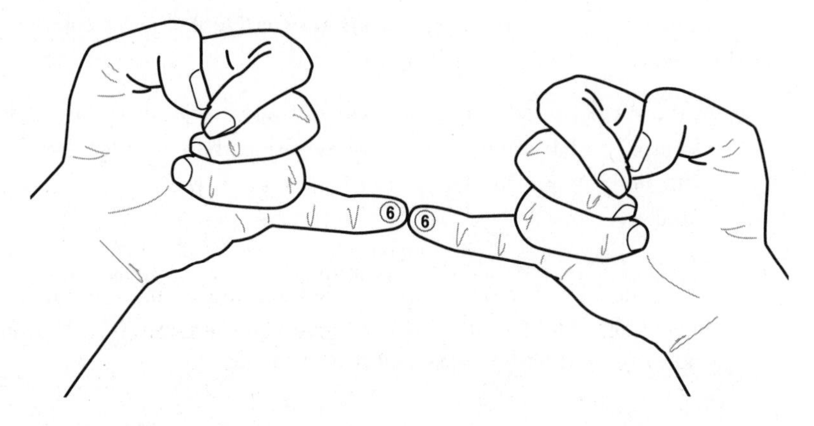

You get the first part of your answer by counting the joined fingers and any fingers below them. In this case, there are no fingers below, so you just count the fingers to get 2. Note that this 2 is actually two tens or 20.

To get the last part, count the fingers above the joined fingers on each hand. You have four on each hand, so you multiply 4 and 4 to get 16.

To arrive at the final answer, add the two numbers. Adding 20 and 16 gives you a final answer of 36.

The 9 Times Table

You may already know the trick to doing your 9 times table. In case you don't, though, I'd like to walk you through the easiest way to do them.

Open both your palms in front of you. Starting with your left thumb, count right, mentally numbering each 1 to 10 as you go.

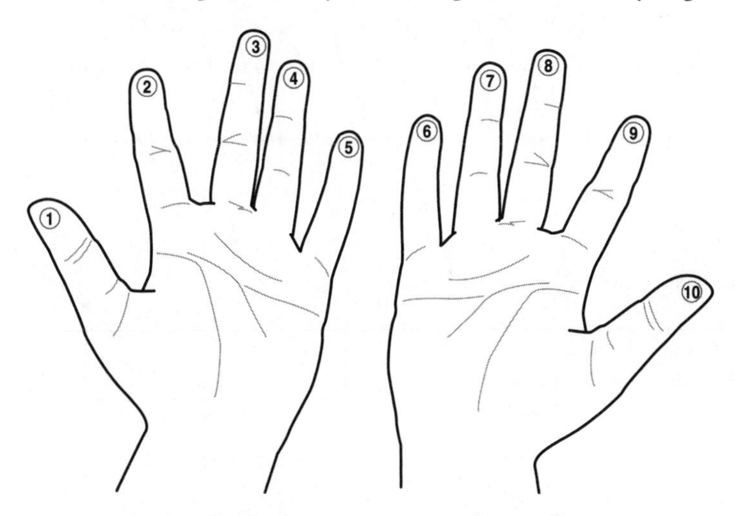

To do the 9 times table, you only need to fold the finger 9 is to be multiplied with. That's all! Let me give you some examples to illustrate how easy this really is.

To solve 9 × 1, you fold the first finger and count the remaining fingers.

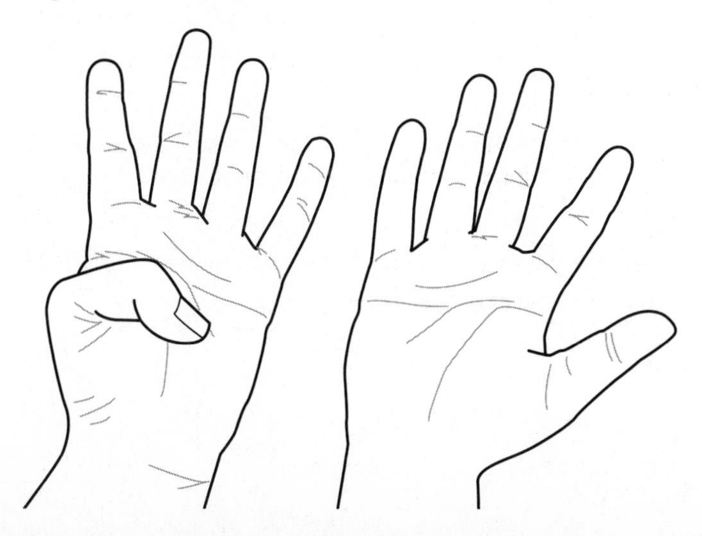

You can see there are nine fingers, so your answer for 9×1 is 9.

Let's take another example, 9×3. Begin by folding the third finger on your left hand.

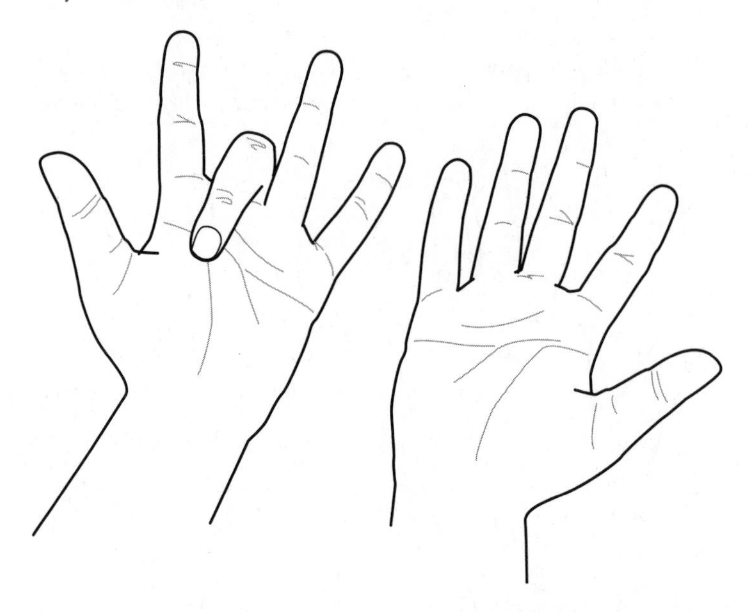

You can see on the left of the folded finger that you have two fingers; on the right of the folded finger, you have seven fingers. Combine 2 and 7 to get your answer, which is 27.

Let's now solve 9×5 on your fingers. Fold the fifth finger. For this one, you can reverse your hands to avoid the discomfort of bending your pinky finger; this way allows you to bend your thumb instead.

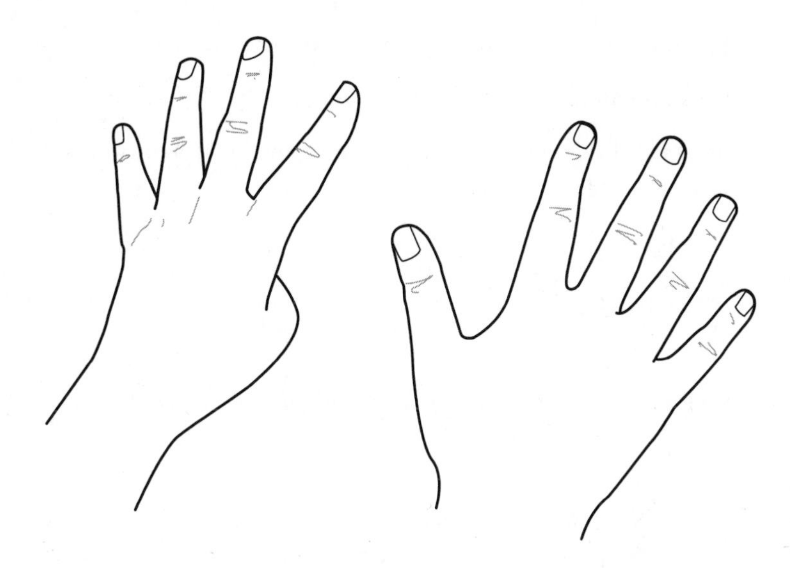

You have four fingers to the left of the folded finger and five fingers to the right of the folded finger. Combined, you get 45 as your answer.

QUICK TIP

If it's too confusing to change it up, you don't have to reverse your hands. You'll still get the same answers—I just like to do it to make the calculations more comfortable.

Special Multiplication by 11

This is a beautiful method that exemplifies what speed math stands for. In a few short steps, you can multiply a number by 11 without reaching for a calculator!

Multiplication of Two-Digit Numbers with 11

To multiply a two-digit number by 11, you split the digits. You then add them to get the middle digit and put everything together to get your answer.

Example 1

Solve the problem 32 × 11.

Step 1: Split the number you're multiplying by 11 into two parts, to make it 3 and 2.

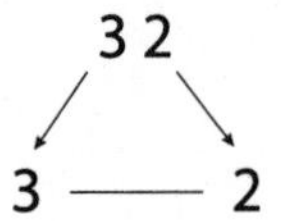

Step 2: Add the numbers: 3 + 2 = 5. Place the 5 in between both the numbers and bring the numbers together.

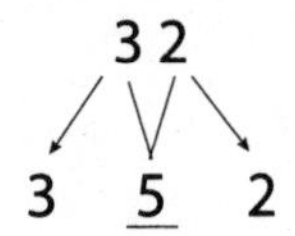

Solution: The answer is 352.

Example 2

Solve the problem 54 × 11.

Step 1: Split 54 into two parts, so 5 becomes the left part and 4 becomes the right part.

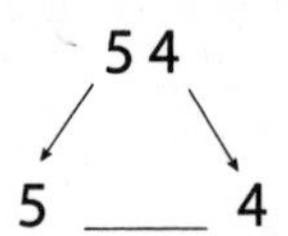

Step 2: Add both parts: 5 + 4 = 9. Place the 9 in the middle and combine the numbers together.

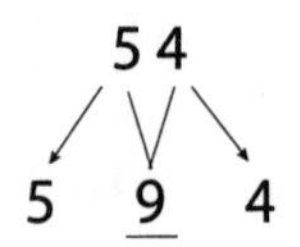

Solution: The answer is 594.

<u>Example 3</u>

Solve the problem 75 × 11.

Step 1: Split 75 into two parts, so 7 goes to the left and 5 to the right.

7 5

7 —— 5

Step 2: Add the two parts: 7 + 5 = 12. Here, because it's a double-digit answer, place the 2 in the middle and carry over the 1 to the left side.

7 5

7 $_1\underline{2}$ 5

Step 3: Add the carryover to the left part: 7 + 1 = 8. Combine all three parts.

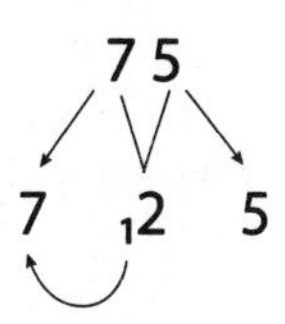

Solution: The answer is 825.

Multiplication of Larger Numbers with 11

To multiply larger numbers by 11, you can use a process called *adding to the neighbor,* which involves adding a digit to the one on its right.

<u>Example 1</u>

Solve the problem 1,234 × 11.

Step 1: Place two lines around the number being multiplied by 11, and add zeroes on the outside of those lines.

0 | 1 2 3 4 | 0

Step 2: Now, begin adding numbers to their neighbor. Start with 4; the neighbor of 4 is 0, so add those digits together: 4 + 0 = 4. This is the ones digit of your answer.

0 | 1 2 3 4 | 0 (4+0 = 4)

4

Step 3: Add 3 to 4 to get the next digit: 3 + 4 = 7.

0 | 1 2 3 4 | 0 (4+0 = 4)
(3+4 = 7)

7 4

Step 4: Add 2 to 3 to get the next digit: 2 + 3 = 5.

0 | 1 2 3 4 | 0 (4+0 = 4)
(3+4 = 7)
(2+3 = 5)

5 7 4

Step 5: Add 1 and 2 to get the next digit: 1 + 2 = 3.

0 | 1 2 3 4 | 0 (4+0 = 4)
(3+4 = 7)
(2+3 = 5)
(1+2 = 3)

3 5 7 4

Step 6: Add the 0 with the 1 to get the final digit: 0 + 1 = 1. Combine the numbers.

0 | 1 2 3 4 | 0

(4+0 = 4)
(3+4 = 7)
(2+3 = 5)
(1+2 = 3)
(0+1 = 1)

1 3 5 7 4

Solution: The answer is 13,574.

<u>Example 2</u>

Solve the problem 782 × 11.

Step 1: Place two lines around 782, and then add two zeroes on the outside of the lines.

0 | 7 8 2 | 0

Step 2: Begin adding numbers to their neighbor. Add 2 to 0 to get the ones digit of your answer: 2 + 0 = 2.

0 | 7 8 2 | 0

2

Step 3: Add 8 and 2 to get the next digit: 8 + 2 = 10. Because it's a double digit, place the 0 below and carry over the 1 to the next step.

0 | 7 8 2 | 0

$_1$0 2

Step 4: Add 7 to 8 to get your next digit: 7 + 8 = 15. You then add the carryover: 15 + 1 = 16. Place the 6 below and carry over the 1 to the next step.

0 | 7 8 2 | 0

$_1$6 0 2

Step 5: Add 0 and 7 to get your final digit: 7 + 0 = 7. Add the carry-over: 7 + 1 = 8. Combine the numbers.

0 | 7 8 2 | 0

8 6 0 2

Solution: The answer is 8,602.

While it may seem like two unnecessary steps, don't skip over adding the first and last numbers to 0 and just put those down as is. Particularly in the case of the last number-and-zero pairing, you may have a carryover from the previous step to include.

<u>Example 3</u>

Solve the problem 5,643 × 11.

Step 1: Place two lines on either side of 5,643 and then add two zeroes outside the lines.

0 | 5 6 4 3 | 0

Step 2: Add neighbors 3 and 0 to get the ones digit: 3 + 0 = 3.

0 | 5 6 4 3 | 0

3

Step 3: Add 4 and 3 to get the next digit: 4 + 3 = 7.

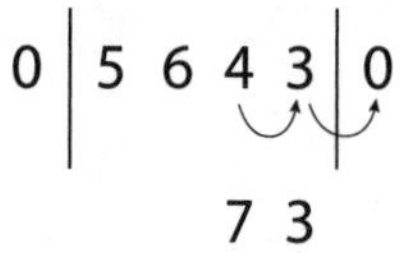

0 | 5 6 4 3 | 0

7 3

Step 4: Add 6 and 4 to get the next digit: 6 + 4 =10. Because it's a double digit, place the 0 below and carry over the 1 to the next step.

0 | 5 6 4 3 | 0

$_1$0 7 3

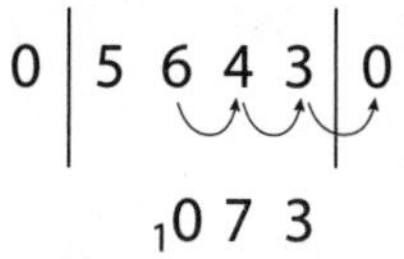

Step 5: Add 5 and 6 to get the next digit: 5 + 6 = 11. Add the carryover from the previous step: 11 + 1 = 12. Place the 2 below and carry over the 1 to the next step.

0 | 5 6 4 3 | 0

$_1$2 0 7 3

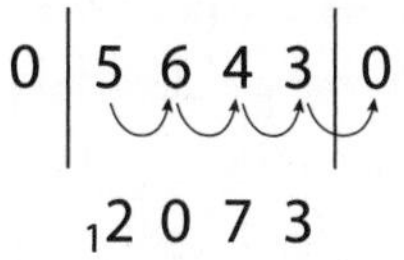

Step 6: Add 0 to 5 to get the final digit: 5 + 0 = 5. Add the carryover: 5 + 1 = 6. Combine the numbers.

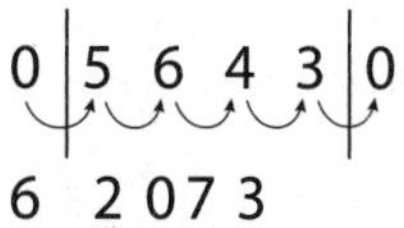

0 | 5 6 4 3 | 0

6 2 07 3

Solution: The answer is 62,073.

Below the Base Method

The base method of speed multiplication is one of the quickest ways to multiply numbers near a base, such as 10 or 100. This method of speed multiplication is perhaps one of the most fascinating and jaw-dropping methods of mental calculation you will ever come across. Through this method, I'll show you that solving small or large multiplication problems doesn't need to be complicated or cumbersome—rather, it can be simple and easy!

There are two important rules to remember while implementing the base method:

1. Add or subtract crosswise.
2. Multiply vertically.

QUICK TIP

This method is used to multiply numbers that are close to powers of 10, which are also called *bases*. For example, 10 and 100 are bases.

I'll begin with some examples of problems with numbers below the base.

Below Base 10

I know you must be thinking, *I already know how to do multiplication problems with numbers under 10; why do I need to use this method to solve it?* I first want to help you build a solid foundation by solving simple sums near to the base 10 before moving on to higher bases.

Example 1

Solve the problem 8 × 7.

Step 1: Figure out how much below 10 each number is. The 8 is 2 less than 10, and the 7 is 3 less than 10; place 2 and 3 to the right. Because both numbers are less than 10, place a minus sign next to them.

8	-2
×7	-3

Step 2: To get the first digit of your answer, subtract crosswise. You can choose either 8 – 3 and 7 – 2, both of which give you 5.

8	-2
×7	-3
5	

Step 3: To get the last digit of your answer, multiply vertically. Here, you multiply -3 × -2, which gives you +6. Bring the two digits together.

8	-2 ↑
×7	-3
5	6

Solution: The answer is 56.

__Example 2__

Solve the problem 9 × 7.

Step 1: The numbers 9 and 7 are both near base 10; 9 is 1 less than 10, and 7 is 3 less than 10, so put 1 and 3 on the right. Because both numbers are less than base 10, place a minus sign next to them.

9	-1
×7	-3

Step 2: Subtract crosswise. You can choose either 9 – 3 or 7 – 1, both of which give you 6. The 6 is the first digit of your answer.

9	-1
×7	-3
6	

Step 3: To get the last digit of your answer, multiply vertically. Here, you multiply -1 × -3, which gives you +3. Bring the two digits together.

9	-1 ↑
×7	-3
6	3

Solution: The answer is 63.

At this point, I want to introduce another basic rule of the base method. Because base 10 has one zero, there will only be one digit on the right side. If the base is 100, there will be two digits on the right side, for base 1,000 there will be three digits on the right side, and so on.

Example 3

Solve the problem 8 × 4.

Step 1: Find the difference of 8 and 4 from 10. Because 8 is less than 10 by 2, put -2 on the right. Because 4 is 6 less than 10, put -6 on the right.

8	-2
×4	-6

Step 2: To get the first digit of your answer, subtract crosswise. You can choose either 8 – 6 or 4 – 2, both of which give you 2.

8	-2
×4	-6
2	

Step 3: To arrive at the second digit of your answer, multiply vertically: -2 × -6 = +12. Because you can't have a double-digit answer on the right, bring down the 2 and carry over the 1 to the left. Add the 1 to the 2 on the left: 1 + 2 = 3. Bring the two digits together.

8	-2
×4	-6
2 ←	12
3	2

Solution: The answer is 32.

Below Base 100

For math problems near base 100, you'll have two digits on the right to account for the two zeroes in 100.

<u>Example 1</u>

Solve the problem 99 × 97.

Step 1: Because 99 and 97 are both near 100, that becomes the base. Like you did for the base 10 problems, find the difference for 99 and 97 from 100. 99 is less than 100 by 1 and 97 is less than 100 by 3. Again, these should be negative, as they're both below the base. Also, because 100 has two zeroes, you need to put two digits on the right. In this case, you add a zero before each digit to make them -01 and -03.

99	-01
× 97	-03

Step 2: Subtract crosswise to get the first part of your answer. You can do 99 – 03 or 97 – 01, both of which give you 96.

99	-01
× 97	-03
96	

Step 3: To arrive at the final part of your answer, multiply vertically: -01 × -03 = +03. Bring the two parts together.

99	-01
× 97	-03
96	03

Solution: The answer is 9,603.

Example 2

Solve the problem 98 × 97.

Step 1: Subtract the numbers from 100. In this case, 98 is less than 100 by 2, so put -02 on the right; 97 is less than 100 by 3, so put -03 on the right.

98	-02
×97	-03

Step 2: Subtract crosswise to get the first part of your answer. You can do either 98 – 03 or 97 – 02, both of which give you 95.

98	-02
×97	-03
95	

Step 3: To get the second part of your answer, multiply vertically: -02 × -03 = +06. Bring the two parts together.

98	-02
×97	-03
95	06

Solution: The answer is 9,506.

Example 3

Solve the problem 88 × 88.

Step 1: Subtract from base 100. In this case, 88 is less than 100 by 12, so put down -12 on the right for both.

88	-12
×88	-12

Step 2: Cross-subtract to get the first part of the answer. Both are exactly the same, so do 88 – 12, which gives you 76.

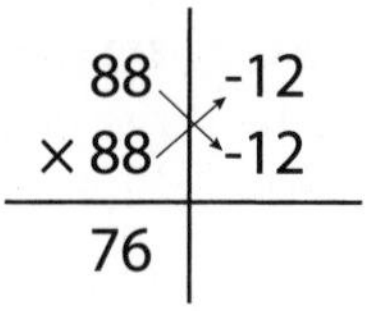

$$\begin{array}{r|l} 88 & -12 \\ \times 88 & -12 \\ \hline 76 & \end{array}$$

Step 3: Multiply vertically to get the second part of your answer: -12 × -12 = +144. Because 144 is a three-digit number, put down 44 with 1 as the carryover.

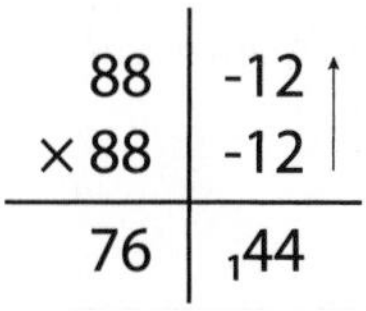

$$\begin{array}{r|l} 88 & -12 \\ \times 88 & -12 \\ \hline 76 & {}_1 44 \end{array}$$

Step 4: Carry the 1 to the left side and add it to 76: 76 + 1 = 77. Bring the two parts together.

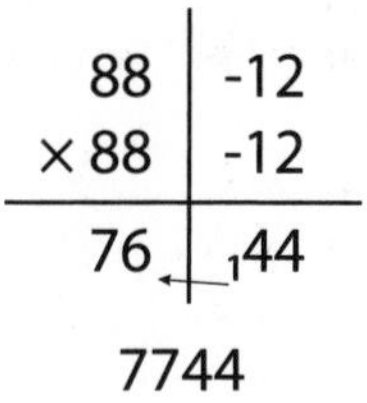

$$\begin{array}{r|l} 88 & -12 \\ \times 88 & -12 \\ \hline 76 & {}_1 44 \end{array}$$

7744

Solution: The answer is 7,744.

Above the Base Method

Now that you know how to solve problems with numbers below the base, let's try some problems in which the numbers are just above the base.

Above Base 10

For numbers in excess of base 10, instead of having a minus sign and subtracting the difference from 10, you have positive numbers which are added crosswise to get the first part of your answer.

Example 1

Solve the problem 11 × 12.

Step 1: Find the difference between the numbers and base 10. In this case, 11 is 1 more and 12 is 2 more than base 10. Write those numbers on the right, with a plus sign next to them.

$$\begin{array}{r|l} 11 & +1 \\ \times 12 & +2 \\ \hline & \end{array}$$

Step 2: Add crosswise to get the first part of the answer. You can do either 11 + 2 or 12 + 1, both of which give you 13.

$$\begin{array}{r|l} 11 & +1 \\ \times 12 & +2 \\ \hline 13 & \end{array}$$

Step 3: Multiply vertically to get the second part of the answer: +1 × +2 = 2. Combine the two parts.

$$\begin{array}{r|l} 11 & +1 \\ \times 12 & +2 \\ \hline 13 & 2 \end{array}$$

Solution: The answer is 132.

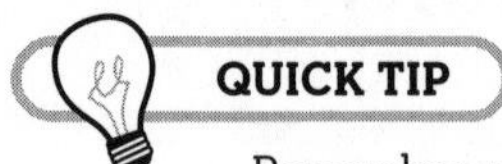

Remember, you can do either addition problem when cross-adding; both should give you the same answer.

<u>Example 2</u>

Solve the problem 12 × 13.

Step 1: Find the difference between the numbers and base 10. In this case, 12 is more than 10 by 2 and 13 is more than 10 by 3, so put +2 and +3 on the right.

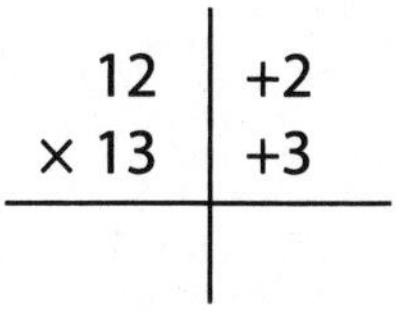

12	+2
× 13	+3

Step 2: Cross-add to get the first part of the answer. You can do either 12 + 3 or 13 + 2, both of which give you 15.

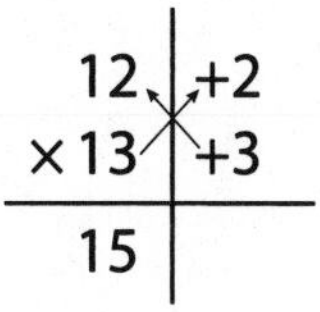

12	+2
× 13	+3
15	

Step 3: To arrive at the second part of the answer, multiply vertically: +2 × +3 = +6. Combine the two parts.

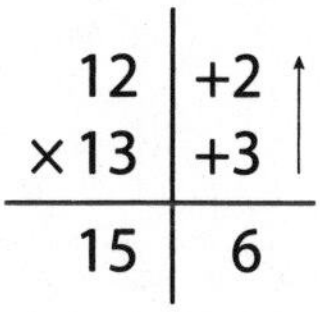

12	+2
× 13	+3
15	6

Solution: The answer is 156.

Example 3

Solve the problem 15×16.

Step 1: Find the difference between the numbers and base 10. In this case, 15 is more than 10 by 5 and 16 is more than 10 by 6, so put +5 and +6 on the right.

$$\begin{array}{r|l} 15 & +5 \\ \times 16 & +6 \\ \hline & \end{array}$$

Step 2: Add crosswise to get the first part of the answer. You can do 15 + 6 or 16 + 5, both of which give you 21.

$$\begin{array}{r|l} 15 & +5 \\ \times 16 & +6 \\ \hline 21 & \end{array}$$

Step 3: Multiply vertically to get the second part of the answer: +5 × +6 = +30. Because you get a two-digit number, put down the 0 and carry over the 3 to the left. Add the carryover: 21 + 3 = 24. Combine the two parts.

$$\begin{array}{r|l} 15 & +5 \\ \times 16 & +6 \\ \hline 21 & {}_{3}0 \end{array}$$

$$240$$

Solution: The answer is 240.

Above Base 100

Let's now take a look at some problems with numbers above base 100 and see how to solve them using the base method.

Example 1

Solve the problem 104×107.

Step 1: Find the difference between the numbers and 100. In this case, 104 is 4 more than 100, and 107 is 7 more than 100. Because there are two zeroes in 100, the numbers on the right have to be two digits; therefore, you put +04 and +07 on the right.

104	+04
× 107	+07

Step 2: Add crosswise to get the first part of the answer. You can do either 104 + 07 or 107 + 04, both of which add to 111.

104	+04
×107	+07
111	

Step 3: To get the second part of the answer, multiply vertically: +04 × +07 = +28. Bring together the two parts.

104	+04
×107	+07
111	28

Solution: The answer is 11,128.

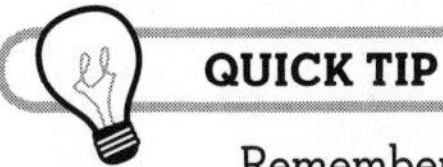

Remember, in problems where the base is 100, there will be two digits on the right side. Any extra digit will be carried over to the left.

Example 2

Solve the problem 105 × 109.

QUICK TIP

For this example, I would like you to visualize the problem in your head and see if you can do it mentally. The answer is 11,445.

Step 1: Find the difference between the numbers and 100. In this case, 105 is 5 more and 107 is 7 more than 100, so put +05 and +07 on the right.

105	+05
×109	+09

Step 2: Add crosswise to get the first part of the answer. You can do 105 + 09 or 109 + 05, both of which equal 114.

105	+05
×109	+09
114	

Step 3: To arrive at the second part of the answer, multiply vertically: +05 × +09 = +45. Combine the two parts.

105	+05
×109	+09
114	45

Solution: The answer is 11,445.

<u>Example 3</u>

Solve the problem 112 × 112.

Step 1: Find the difference between the numbers and 100. In this case, 112 is more than 100 by 12, so put +12 for both numbers on the right.

112	+12
×112	+12

Step 2: Add crosswise to get the first part of the answer. Because both addition problems are the same, you simply add 112 and 12, which gives you 124.

112	+12
×112	+12
124	

Step 3: To get the second part of the answer, multiply vertically: +12 × +12 = 144. Because it's a three-digit number, put down the 44 and carry over the 1 to the left. Add the carryover to 124: 124 + 1 = 125. Combine the two parts.

112	+12
×112	+12
124	144

12544

Solution: The answer is 12,544.

Above and Below the Base Method

You've seen that numbers can be below the base or above the base. But sometimes, you'll multiply numbers in which one is below the base and the other is above the base. What should you do in such cases? The following takes you through how to solve them depending on the base.

Above and Below Base 10

When you have one number above base 10 and one below base 10, you'll be dealing with a negative number. To get rid of it, you multiply the left side by 10 and subtract the right side from it.

<u>Example 1</u>

Solve the problem 11 × 9.

Step 1: Find the difference between the numbers and 10. In this case, 11 is 1 more than 10, so put down +1 on the right; 9 is 1 less than 10, so put -1 on the right.

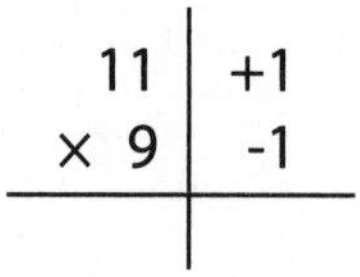

11	+1
× 9	-1

Step 2: Add or subtract crosswise to get the first part of your answer. You can do 11 – 1 or 9 + 1, both of which equal 10. Next, multiply vertically to get the second part: +1 × -1 = -1.

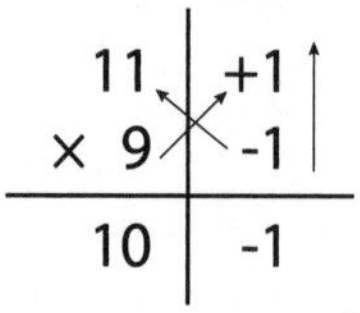

11	+1
× 9	-1
10	-1

Step 3: Because your answer can't be negative, you multiply the left side by 10, which is the base: 10 × 10 = 100. You then subtract the left side from the right side.

11	+1 ↑
× 9	-1
10	-1
×10	

100 - 1 = 99

Solution: The answer is 99.

QUICK TIP

You may be wondering, *Why do I have to do a simple problem like 11 × 9 this way?* These examples are simply building the foundation to deal with larger numbers. When the larger numbers come, you'll be able to deal with them smoothly.

<u>**Example 2**</u>

Solve the problem 14 × 8.

Step 1: Find the difference between the numbers and 10. In this case, 14 is 4 more than 10, so you put +4 on the right; 8 is 2 less than 10, so put -2 on the right.

14	+4
× 8	-2

Step 2: Add or subtract crosswise to get the first part of your answer. You can do 14 – 2 or 8 + 4, both of which give you 12. To get the second part of the answer, multiply vertically: +4 × -2 = -8.

14	+4 ↑
× 8	-2
12	-8

Step 3: Your base is 10, so multiply the left side by 10: 12 × 10 = 120. Subtract the left side from the right side.

14	+4
× 8	-2
12	-8
×10	

120 - 8 = 112

Solution: The answer is 112.

<u>Example 3</u>

Solve the problem 17 × 8.

Step 1: Find the difference between the numbers and 10. In this case, 17 is 7 more than 10, so put +7 on the right; 8 is 2 less than 10, so put -2 on the right.

17	+7
× 8	-2

Step 2: Add or subtract crosswise to get the first part of the answer. You can do 17 – 2 or 8 + 7, both of which equal 15. Multiply vertically to get the second part: +7 × -2 = -14.

17	+7
× 8	-2
15	-14

Step 3: Because the base is 10, multiply the left side by 10: 15 × 10 = 150. Subtract the left side from the right side.

17	+7
× 8	-2
15	-14
×10	

150 -14 =136

Solution: The answer is 136.

Above and Below Base 100

The process for solving multiplication problems with numbers above and below base 100 is pretty much the same as what you did for base 10. The only difference is that you multiply by 100 instead of 10.

<u>Example 1</u>

Solve the problem 106 × 92.

Step 1: Find the difference between the numbers and 100. For 106, the excess is 6 and for 92, the deficiency is 8. Therefore, put down +6 and -8 on the right.

$$\begin{array}{r|l} 106 & +6 \\ \times\ 92 & -8 \\ \hline & \end{array}$$

Step 2: Add or subtract crosswise to get the first part of the answer. You can do 106 – 8 or 92 + 6, both of which equal 98. So 98 is our first part. Multiply vertically to get the second part: 6 × -8 = -48.

$$\begin{array}{r|l} 106 & +6 \\ \times\ 92 & -8 \\ \hline 98 & -48 \end{array}$$

Step 3: Multiply the left side by 100: 98 × 100 = 9,800. Subtract the left side from the right side.

$$\begin{array}{r|l} 106 & +6 \\ \times\ 92 & -8 \\ \hline 98 & -48 \\ \times 100 & \\ \hline \multicolumn{2}{c}{9800 - 48 = 9752} \end{array}$$

Solution: The answer is 9,752.

Example 2

Solve the problem 105 × 88.

Step 1: Find the difference between the numbers and 100. In this case, 105 is in excess 100 by 5, so put +5 on the right; 88 is below hundred by 12, so put -12 on the right.

105	+5
× 88	-12

Step 2: Add or subtract crosswise to get the first part of the answer. You can do 105 – 12 or 88 + 5, both of which give you 93. Multiply vertically to get the second part: +5 × -12 = -60.

105	+5
× 88	-12
93	-60

Step 3: Multiply the left hand side by 100: 93 × 100 = 9,300. Subtract the left side from the right side.

105	+5
× 88	-12
93	-60
×100	

9300 - 60 = 9240

Solution: The answer is 9,240.

Example 3

Solve the problem 117×91.

Step 1: Find the difference between the numbers and 100. In this case, 117 is 17 more than 100, so put +17 on the right; 91 is 9 less than 100, so put -9 on the right.

$$\begin{array}{r|l} 117 & +17 \\ \times\ 91 & -9 \\ \hline & \end{array}$$

Step 2: Add or subtract crosswise to get the first part of the answer. You can do $117 - 9$ or $91 + 17$, both of which equal 108. Multiply vertically to get the second part: $+17 \times -9 = -153$.

$$\begin{array}{r|l} 117 & +17 \\ \times\ 91 & -9 \\ \hline 108 & -153 \end{array}$$

Step 3: Multiply the left side by base 100: $108 \times 100 = 10{,}800$. Subtract the left side from the right side.

$$\begin{array}{r|l} 117 & +17 \\ \times\ 91 & -9 \\ \hline 108 & -153 \\ \times 100 & \\ \hline \multicolumn{2}{l}{10800 - 153 = 10647} \end{array}$$

Solution: The answer is 10,647.

Multiples and Submultiples

If you have to multiply numbers that are both far lower or far higher than the base, you can use a base that's a multiple of 10. The following give you some examples of this.

<u>Example 1</u>

Solve the problem 46 × 48.

Step 1: For 46 and 48, instead of using base 100, make the base 50 and find the difference. This gives you -4 and -2, which you put on the right side.

46	-4
× 48	-2

Step 2: Cross-subtract to get the first part of your answer. You can do 46 – 2 or 48 – 4, both of which equal 44. Multiply vertically to get the second part of your answer: -4 × -2 = +8.

46	-4
× 48	-2
44	8

Step 3: Because the base is 50, which you get by multiplying 10 × 5, multiply the left side by 5: 44 × 5 = 220. Bring the two parts together.

46	-4
× 48	-2
44	8
× 5	
2208	

Solution: The answer is 2,208.

Example 2

Solve the problem 59 × 59.

Step 1: Here, the base could be 50 or 60. However, because both numbers are closer to 60, go ahead and use 60 as the base. Find the difference between the numbers and 60. Both are 1 less than 60, so put -1 on the right.

59	-1
× 59	-1

Step 2: Subtract crosswise for the first part of the answer. Both are 59 – 1, which equals 58. Multiply vertically to get the second part of the answer: -1 × -1 = 1.

59	-1
× 59	-1
58	1

Step 3: Because the base is 60, multiply the left side by 6: 58 × 6 = 348. Bring the two parts together.

59	-1
× 59	-1
58	1
× 6	
3481	

Solution: The answer is 3,481.

Example 3

Solve the problem 216 × 204.

Step 1: Both are close to 200, so make 200 the base and find the difference. Because 216 is 16 more and 204 is 4 more than 200, put +16 and +04 on the right.

216	+16
× 204	+04

Step 2: Add crosswise to get the first part of the answer. You can do 216 + 04 or 204 + 16, both of which equal 220. Multiply vertically to get the second part: +16 × +04 = +64.

216	+16
× 204	+04
220	64

Step 3: Because 100 × 2 = 200, multiply the left side by 2: 220 × 2 = 440. Bring the two parts together.

216	+16
× 204	+04
220	64
× 2	
44064	

Solution: The answer is 44,064.

Don't multiply the left side by the actual base number. Instead, use the number you have to multiply with to get to base 10 or base 100. For example, if your base is 300, you don't multiply the left by 300; instead, you multiply by 3, because the base is 100 × 3.

Combined Problems

If you have a problem in which one number is far below the base, and the other is far above it, you can use a multiple of 10 as the base. As you'll see in the following examples, this process mimics the above and below the base method.

<u>Example 1</u>

Solve the problem 199 × 202.

Step 1: Make the base 200 for these numbers and find the difference. Because 199 is 1 less and 202 is 2 more than 200, put -01 and +02 on the right.

$$\begin{array}{r|r} 199 & -01 \\ \times\ 202 & +02 \\ \hline & \end{array}$$

Step 2: Add or subtract crosswise to get the left side. You can do 199 + 02 or 202 – 01, both of which equal 201. Multiply vertically to get the right side: +02 × -01 = -02.

$$\begin{array}{r|r} 199 & -01 \\ \times\ 202 & +02 \\ \hline 201 & -02 \end{array}$$

Step 3: Because the base is 200, multiply the left side by 2: 201 × 2 = 402. Multiply 402 by 100: 402 × 100 = 40,200.

$$\begin{array}{r|r} 199 & -01 \\ \times\ 202 & +02 \\ \hline 201 & -02 \\ \times\ 2 & \\ \hline 402 & \\ \times\ 100 & \\ \hline 40200 & \end{array}$$

Step 4: Subtract the right side from the left side.

199	-01
× 202	+02
201	-02

```
  201
  x 2
  ---
  402
× 100
 = 40200 - 02
 = 40198
```

Solution: The answer is 40,198.

<u>**Example 2**</u>

Solve the problem 297 × 304.

Step 1: Make the base 300 and find the difference. In this case, 297 is 3 less and 304 is 4 more than 300, so put -03 +04 on the right side.

297	-03
× 304	+04

Step 2: Add or subtract crosswise for the left side. You can do 297 + 04 or 304 – 03, both of which equal 301. Multiply vertically for the right side: +04 × -03 = -12.

297	-03
× 304	+04
301	-12

Step 3: Because the base is 300, multiply the left side by 3: $301 \times 3 = 903$. Multiply 903 by 100: $903 \times 100 = 90{,}300$.

$$\begin{array}{r|r} 297 & -03 \\ \times\, 304 & +04 \\ \hline 301 & -12 \\ \times\, 3 & \\ \hline 903 & \end{array}$$

Step 4: Subtract the right side from the left side.

$$\begin{array}{r|r} 297 & -03 \\ \times\ \ 304 & +04 \\ \hline 301 & -12 \\ \times\, 3 & \\ \hline 903 & \\ \times\ 100 & \\ \hline \end{array}$$

$$= 90300 - 12$$
$$= 90288$$

Solution: The answer is 90,288.

SPEED BUMP

Unlike in earlier problems involving the base method, simply combining the two parts of the answer won't yield the correct response. You have to subtract the right side from the left side to get your answer.

General Multiplication Using the Vertically and Crosswise Method

The base method is great if both numbers are close to the same base, but you'll encounter many problems in which the numbers aren't even remotely close to each other. Enter the vertically and crosswise method.

As the name indicates, the vertically and crosswise method uses a vertical and crosswise pattern to multiply numbers. This method is extremely versatile, with varied applications (as you'll see throughout the book).

Two-Digit by Two-Digit Multiplication

To solve any problems in which a two-digit number is multiplied by a two-digit number, use the following version of the vertically and crosswise method.

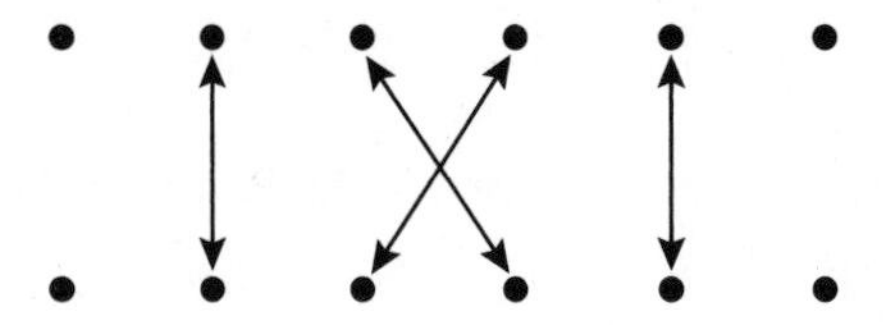

You multiply vertically, then crosswise, and vertically again. The following examples show you how to apply this method.

<u>Example 1</u>

Solve the problem 12 × 43.

Step 1: To get the first digit of your answer, multiply vertically on the right side: 3 × 2 = 6.

$$\begin{array}{r} 12 \\ \times 43 \\ \hline 6 \end{array}$$

Step 2: To get the next digit, multiply crosswise and add: $(3 \times 1) + (4 \times 2) = 3 + 8 = 11$. Put down 1 as the tens-place digit and carry over the 1 to the next step.

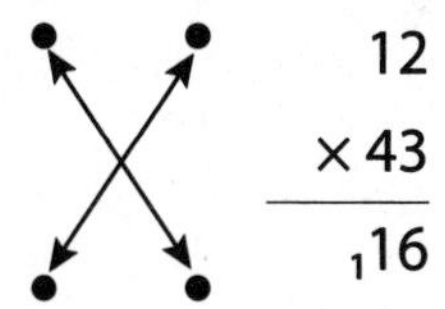

Step 3: To get the last digit, multiply vertically on the left side: $4 \times 1 = 4$. Add the carryover: $4 + 1 = 5$.

$$\begin{array}{r} 12 \\ \times 43 \\ \hline 516 \end{array}$$

Solution: The answer is 516.

<u>Example 2</u>

Solve the problem 78×69.

Step 1: Multiply vertically on the right: $9 \times 8 = 72$. Put down 2 and carry over the 7 to the next step.

$$\begin{array}{r} 78 \\ \times 69 \\ \hline {}_{7}2 \end{array}$$

Step 2: Multiply crosswise and add; include the carryover 7 in the addition: $(9 \times 7) + (6 \times 8) + 7 = 63 + 48 + 7 = 118$. Put down 8 in the tens place and carry over the 11 to the next step.

$$\begin{array}{r} 78 \\ \times 69 \\ \hline {}_{11}82 \end{array}$$

Step 3: Multiply vertically on the right and add the carryover: (6 × 7) + 11 = 42 + 11 = 53.

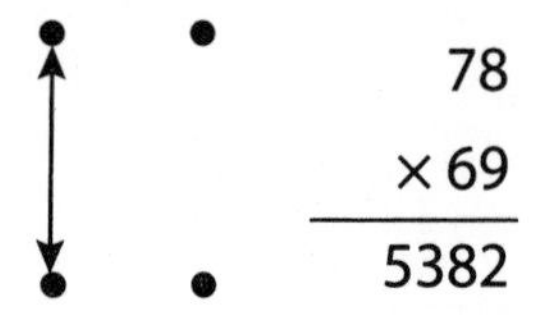

Solution: The answer is 5,382.

QUICK TIP

The examples have shown how to do the vertically and crosswise method from right to left, but you can also do it from left to right. If you have any carryovers, you add them to the previous step rather than the next step.

Example 3

Solve the problem 34 × 8.

Step 1: Because 8 isn't a two-digit number, put a zero before the 8 before starting the method.

$$\begin{array}{r} 34 \\ \times\, 08 \\ \hline \end{array}$$

Step 2: Multiply vertically on the right: 8 × 4 = 32. Put down the 2 and carry over the 3 to the next step.

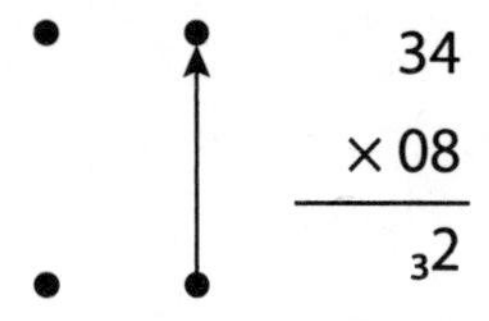

Step 3: Multiply crosswise and add; don't forget to include the carryover 3 in the addition: (8 × 3) + (4 × 0) + 7 = 24 + 0 + 3 = 27. Put 7 down in the tens place and carry over the 2 to the next step.

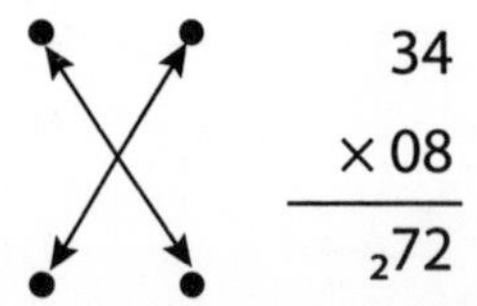

Step 4: Multiply vertically on the right side and add the carryover: $(0 \times 3) + 2 = 0 + 2 = 2$.

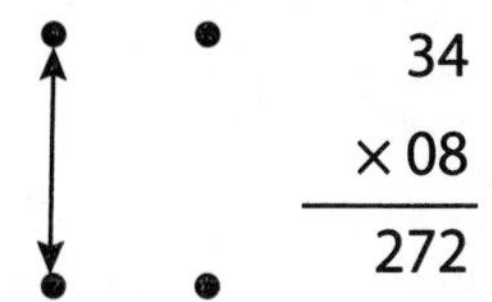

Solution: The answer is 272.

Three-Digit by Three-Digit Multiplication

This is an extension of the two-digit version of the vertically and crosswise method. The three-digit version of the method involves five simple steps.

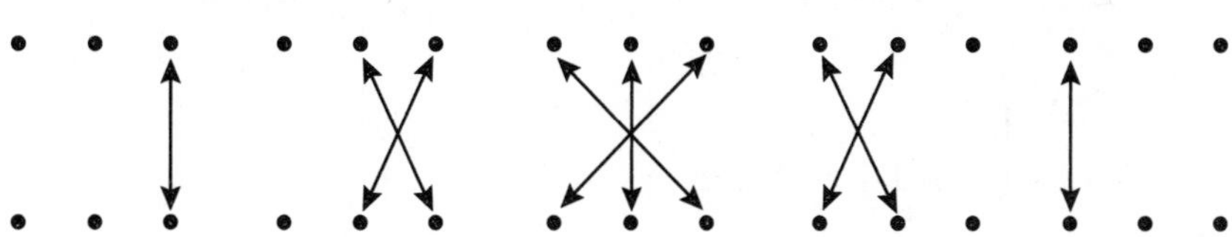

Think of the steps as vertical, crosswise, star, crosswise, and vertical. The following examples take you through the process for multiplying a three-digit number by another three-digit number.

<u>Example 1</u>

Solve the problem 123×456.

Step 1: Multiply vertically on the right side: $6 \times 3 = 18$. Put 8 down in the units place figure and carry over the 1 to the next step.

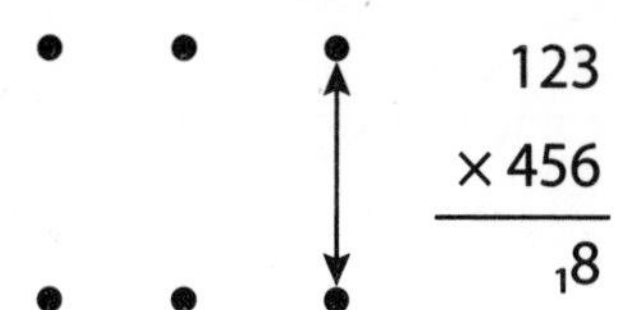

Step 2: Multiply the right and middle columns crosswise and add; don't forget to include the carryover in the addition: (6 × 2) + (5 × 3) + 1 = 12 + 15 + 1 = 28. Put 8 down in the tens place and carry over the 2 to the next step.

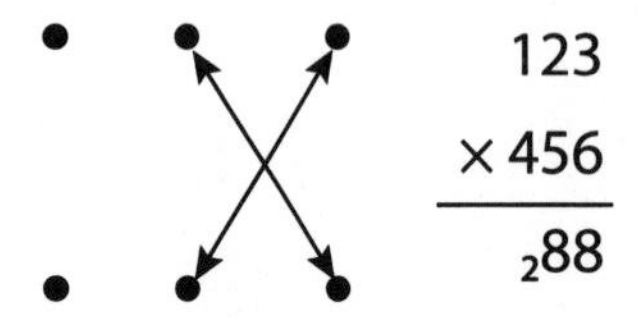

Step 3: Multiply in the star pattern—the bottom-right and top-left numbers, the top and bottom middle numbers, and the bottom-left and top-right numbers—and add; don't forget to include the carryover in the addition: (6 × 1) + (4 × 3) + (5 × 2) + 2 = 6 + 12 + 10 + 2 = 30. Put down 0 and carry over the 3 to the next step.

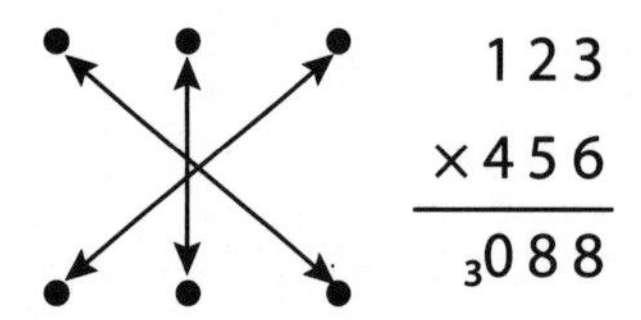

Step 4: Multiply the left and middle columns crosswise and add; don't forget to include the carryover in the addition: (5 × 1) + (4 × 2) + 3 = 5 + 8 + 3 = 16. Put down 6 and carry over the 1 to the next step.

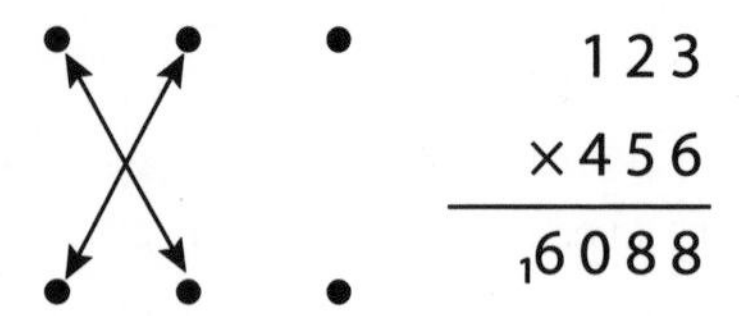

Step 5: Multiply vertically on the left side: 4 × 1 = 4. Add the carryover: 4 + 1 = 5.

$$\begin{array}{r} 123 \\ \times 456 \\ \hline 56088 \end{array}$$

Solution: The answer is 56,088.

Example 2

Solve the problem 231 × 745.

Step 1: Multiply vertically on the right side: 5 × 1 = 5.

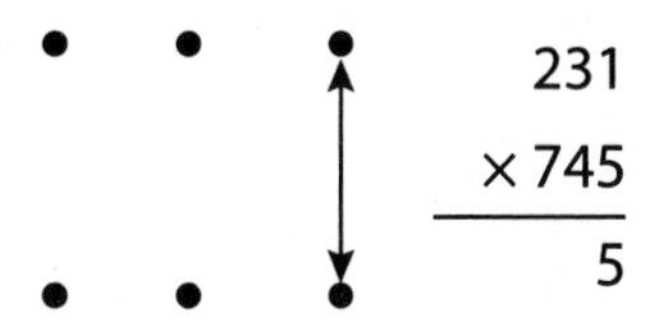

Step 2: Multiply the right and middle columns crosswise and add: (5 × 3) + (4 × 1) = 15 + 4 = 19. Put down 9 in the tens place and carry over the 1 to the next step.

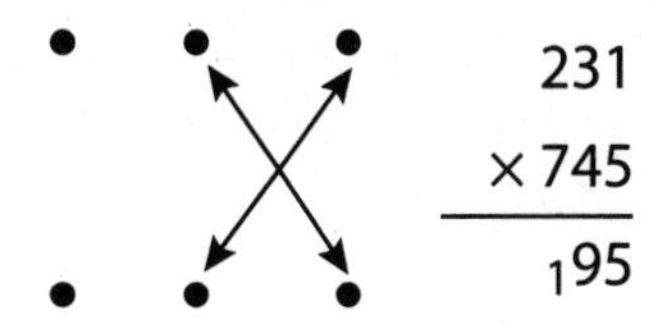

Step 3: Multiply in the star pattern—the bottom-right and top-left numbers, the top and bottom middle numbers, and the bottom-left and top-right numbers—and add; don't forget to include the carryover in the addition: (5 × 2) + (7 × 1) + (4 × 3) + 1 = 10 + 7 + 12 + 1 = 30. Put 0 down and carry over the 3 to the next step.

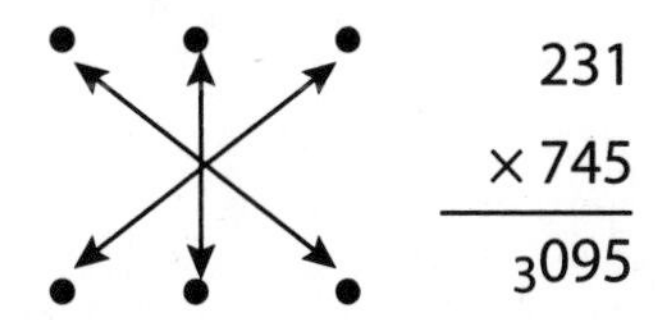

Step 4: Multiply the left and middle columns crosswise and add; don't forget to include the carryover in the addition: (4 × 2) + (7 × 3) + 3 = 8 + 21 + 3 = 32. Put down 2 in the thousands place and carry over the 3 to the next step.

231
× 745
32095

Step 5: Multiply vertically on the left side: 7 × 2 = 14. Add the carryover: 14 + 3 = 17.

$$\begin{array}{r} 231 \\ \times\, 745 \\ \hline 172095 \end{array}$$

Solution: The answer is 172,095.

QUICK TIP

Take some time to learn the star in particular in the three-digit version. In the beginning, you may make some errors here and there, but ultimately I'm sure you'll increase your speed and accuracy in solving three-digit multiplication problems.

__Example 3__

Solve the problem 321 × 042.

Step 1: Multiply vertically on the right side: 2 × 1 = 2. Put 2 down in the units place.

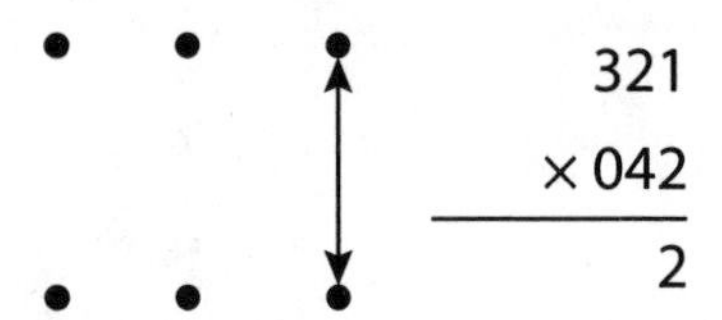

$$\begin{array}{r} 321 \\ \times\, 042 \\ \hline 2 \end{array}$$

Step 2: Multiply the right and middle columns crosswise and add: (2 × 2) + (4 × 1) = 4 + 4 = 8. Put down 8 in the tens place.

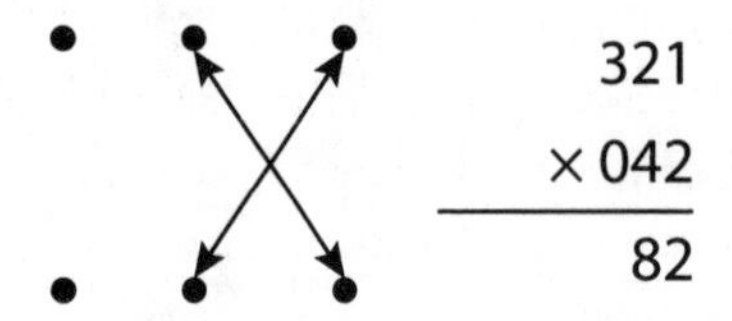

$$\begin{array}{r} 321 \\ \times\, 042 \\ \hline 82 \end{array}$$

Step 3: Multiply in the star pattern—the bottom-right and top-left numbers, the top and bottom middle numbers, and the bottom-left and top-right numbers—and add: (2 × 3) + (0 × 1) + (4 × 2) = 6 + 0 + 8 = 14. Put 4 in the thousands place and carry over the 1 to the next step.

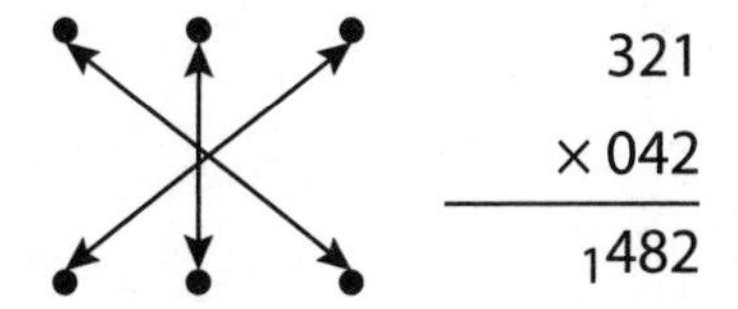

Step 4: Multiply the left and middle columns crosswise and add; don't forget to include the carryover in the addition: (4 × 3) + (0 × 2) + 1 = 12 + 0 + 1 = 13. Put down 3 in the thousands place and carry over the 1 to the next step.

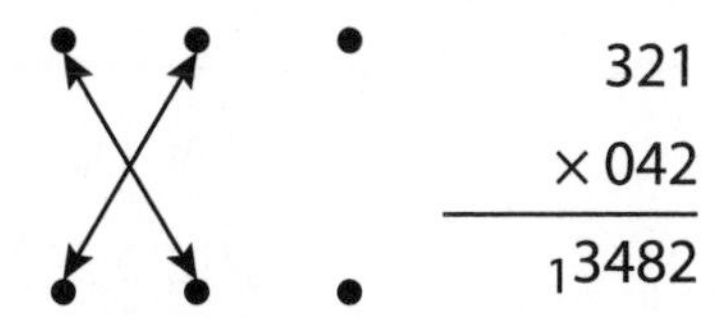

Step 5: Multiply vertically on the left side: 0 × 3 = 0. Add the carryover: 0 + 1 = 1.

321
× 042
13482

Solution: The answer is 13,482.

The Least You Need to Know

- You can use your fingers to do your times tables, whether it's by counting the joints of the fingers or joining two and counting the digits above and below them.
- When multiplying a two-digit number by 11, split the two digits and add to get the middle digit.
- If your number is near a base, you can round to the base to help you find the answer.
- Depending on how many digits they have, you can use a vertical and crosswise pattern to multiply two numbers.

CHAPTER

2

Addition

In This Chapter

- Adding left to right
- Calculating addition problems with numbers near 10 or a multiple of 10
- Solving addition problems with number splitting

Addition is one of the first operations you learned as a child. And if you're like me, you loved doing our 2 + 2's. In this chapter, you'll learn some simple addition methods that will make the process once again feel like child's play. Once you know them, you'll be able to do addition problems faster and without the aid of a calculator or even a pencil and paper.

Left-to-Right Addition

You can speed up your addition by adding the numbers in each column from left to right. In the following sections, I take you through how to use this method on two-digit and three-digit numbers.

Two-Digit Numbers

For two-digit numbers, you first add the left column of numbers, followed by the right column of numbers. You then combine the digits in the middle, which gives you your answer.

Example 1

Solve the problem 78 + 45.

Step 1: Add the numbers in the first column: 7 + 4 = 11.

$$\begin{array}{r} 78 \\ +45 \\ \hline 11 \end{array}$$

Step 2: Add the numbers in the second column: 8 + 5 = 13.

$$\begin{array}{r} 78 \\ +45 \\ \hline 11,13 \end{array}$$

Step 3: Combine the middle digits: 1 + 1 = 2.

$$\begin{array}{r} 78 \\ +45 \\ \hline 11,13 \\ 123 \end{array}$$

Solution: The answer is 123.

Make sure you're not merely writing these examples down or plugging them into your calculator. Practice doing them in your head.

Example 2

Solve the problem 87 + 69.

Step 1: Add the numbers in the first column: 8 + 6 = 14.

$$\begin{array}{r} 87 \\ +69 \\ \hline 14 \end{array}$$

Step 2: Add the numbers in the second column: 7 + 9 = 16.

$$\begin{array}{r} 87 \\ +69 \\ \hline 14,16 \end{array}$$

Step 3: Combine the middle digits: 4 + 1 = 5.

$$\begin{array}{r} 87 \\ +69 \\ \hline 14,16 \\ 156 \end{array}$$

Solution: The answer is 156.

Example 3

Solve the problem 48 + 97.

Step 1: Add the numbers in the first column: 4 + 9 = 13.

$$\begin{array}{r} 48 \\ +97 \\ \hline 13 \end{array}$$

Step 2: Add the numbers in the second column: 8 + 7 = 15.

$$\begin{array}{r} 48 \\ +97 \\ \hline 13,15 \end{array}$$

Step 3: Combine the middle digits: 1 + 3 = 4.

$$\begin{array}{r} 48 \\ +97 \\ \hline 13,15 \\ 145 \end{array}$$

Solution: The answer is 145.

Three-Digit Numbers

You can build on what you learned doing two-digit addition left to right by tackling three-digit addition. This involves another step of combining numbers in the middle.

Example 1

Solve the problem 582 + 759.

Step 1: Add the numbers in the first column: 5 + 7 = 12.

$$\begin{array}{r} 582 \\ +759 \\ \hline 12 \end{array}$$

Step 2: Add the numbers in the second column: 8 + 5 = 13. Combine the middle digits: 2 + 1 = 3. This gives you a value of 133.

$$\begin{array}{r} 582 \\ +759 \\ \hline 12,13 \\ 133 \end{array}$$

Step 3: Add the numbers in the third column: 2 + 9 = 11. Combine the middle digits for the values 133 and 11: 1 + 3 = 4.

$$\begin{array}{r} 582 \\ +759 \\ \hline 12,13 \\ 133,11 \\ 1341 \end{array}$$

Solution: The answer is 1,341.

<u>Example 2</u>

Solve the problem 834 + 786.

Step 1: Add the numbers in the first column: 8 + 7 = 15.

$$\begin{array}{r} 834 \\ +786 \\ \hline 15 \end{array}$$

Step 2: Add the numbers in the second column: 3 + 8 = 15. Combine the middle digits: 5 + 1 = 6. This gives you a value of 161.

$$\begin{array}{r} 834 \\ +786 \\ \hline 15,11 \\ 161 \end{array}$$

Step 3: Add the numbers in the third column: 4 + 6 = 10. Combine the middle digits for the values 161 and 10: 1 + 1 = 2.

$$\begin{array}{r} 834 \\ +786 \\ \hline 15,11 \\ 161,10 \\ 1620 \end{array}$$

Solution: The answer is 1,620.

QUICK TIP

It probably feels strange to add left to right instead of right to left, like most schools teach. However, you can probably see after doing a few problems how much faster it is not just marking off all the carryovers and then doing the addition. You're dealing with the numbers individually and combining as needed, which is a better use of your time and energy.

Example 3

Solve the problem 983 + 694.

Step 1: Add the numbers in the first column: 9 + 6 = 15.

$$\begin{array}{r} 983 \\ +\,694 \\ \hline 15 \end{array}$$

Step 2: Add the numbers in the second column: 8 + 9 = 17. Combine the middle digits: 5 + 1 = 6. This gives you a value of 167.

$$\begin{array}{r} 983 \\ +\,694 \\ \hline 15,17 \\ 167 \end{array}$$

Step 3: Add the numbers in the third column: 3 + 4 = 7. Because this is a single digit, you simply bring down the 7 to the end of the answer—no addition is necessary.

$$\begin{array}{r} 983 \\ +\,694 \\ \hline 15,17 \\ 167,7 \\ 1677 \end{array}$$

Solution: The answer is 1,677.

Rapid Left-to-Right Columnar Addition

When doing addition, you won't always be adding two short numbers. Sometimes, you'll have to have many multiple-digit numbers at the same time. Using left-to-right addition, I'll show you how to do it quickly and easily.

Example 1

Solve the problem 5,273 + 7,372 + 6,371 + 9,782.

Step 1: Add the numbers in the first column: 5 + 7 + 6 + 9 = 27.

```
  5273
  7372
  6371
+ 9782
------
 27
```

Step 2: Add the numbers in the second column: 2 + 3 + 3 + 7 = 15. Combine the middle digits: 7 + 1 = 8. This gives you a value of 285.

```
  5273
  7372
  6371
+ 9782
------
 27,15
  ↷
 285
```

Step 3: Add the numbers in the third column: 7 + 7 + 7 + 8 = 29. Combine the middle digits for the values 285 and 29: 5 + 2 = 7. This gives you a value of 2,879.

```
  5273
  7372
  6371
+ 9782
------
 27,15
  ↷
 285,29
    ↷
 2879
```

Step 4: Add the numbers in the fourth column: 3 + 2 + 1 + 2 = 8. Because this is a single-digit number, you bring it down to the end of the answer.

```
  5273
  7372
  6371
+ 9782
------
 27,15
 285,29  8
 28798
```

Solution: The answer is 28,798.

SPEED BUMP

Remember, because you're bringing down a single digit, there will be no combining and adding. The answer will simply be 28,798.

Example 2

Solve the problem 8,336 + 4,283 + 3,428+ 9,373.

Step 1: Add the numbers in the first column: 8 + 4 + 3 + 9 = 24.

```
  8336
  4283
  3428
+ 9373
------
 24
```

Step 2: Add the numbers in the second column: 3 + 2 + 4 + 3 = 12. Combine the middle digits: 4 + 1 = 5. This gives you a value of 252.

```
  8336
  4283
  3428
+ 9373
------
 24,12
 252
```

Step 3: Add the numbers in the third column: 3 + 8 + 2 + 7 = 20. Combine the middle digits for the values 252 and 20: 2 + 2 = 4. This gives you a value of 2,540.

```
  8336
  4283
  3428
+ 9373
------
 24,12
  252,20
   2540
```

Step 4: Add the numbers in the fourth column: 6 + 3 + 8 + 3 = 20. Combine the middle digits for the values 2,540 and 20: 0 + 2 = 2.

```
  8336
  4283
  3428
+ 9373
------
 24,12
  252,20
   2540,20
    25420
```

Solution: The answer is 25,420.

Addition with Numbers Near 10 or a Multiple of 10

Numbers near 10 or a multiple of 10—such as 9, 18, 27, 36, and so on—are very simple to add. To simplify addition problems which include these numbers, all you have to do is make the digit 10 or the multiple of 10 it's closest to and then subtract how much you added in after you get your answer. These require no carryover and are possible to do just mentally!

Example 1

Solve the problem 24 + 9.

Step 1: Change the 9 to a 10 by adding 1.

9 + 1 = 10

Step 2: Add the numbers.

24 + 10 = 34

Step 3: Because you had to add 1 to make the 9 a 10, you must now subtract 1 from the sum.

34 – 1 = 33

Solution: The answer is 33.

QUICK TIP

Don't forget to subtract how much you had to add in to make the digit 10 or a multiple of 10.

Example 2

Solve the problem 46 + 18.

Step 1: Change the 18 to a 20 by adding 2.

18 + 2 = 20

Step 2: Add the numbers.

46 + 20 = 66

Step 3: Subtract 2 from the sum.

66 – 2 = 64

Solution: The answer is 64.

Example 3

Solve the problem 458 + 38.

Step 1: Change the 38 to a 40 by adding 2.

38 + 2 = 40

Step 2: Add the numbers.

$$458 + 40 = 498$$

Step 3: Subtract 2 from the sum.

$$498 - 2 = 496$$

Solution: The answer is 496.

Number Splitting

Number splitting is a very useful method that allows you to split a big problem into two or three smaller parts. It reduces the number of steps you need to take to compute the answer.

The following examples may look slightly difficult because you're adding two four-digit numbers. However, with number splitting, finding the answers is a piece of cake!

<u>**Example 1**</u>

Solve the problem 4,381 + 2,707.

Step 1: Split the problem into two parts by drawing or imagining a line down the middle.

$$\begin{array}{r|l} 43 & 81 \\ +27 & 07 \\ \hline & \end{array}$$

Step 2: Add the numbers on the left side of the split: 43 + 27 = 70.

$$\begin{array}{r|l} 43 & 81 \\ +27 & 07 \\ \hline 70 & \end{array}$$

Step 3: Add the numbers on the right side of the split: 81 + 07 = 88. Bring the two parts together.

$$\begin{array}{r|l} 43 & 81 \\ +27 & 07 \\ \hline 70 & 88 \end{array}$$

Solution: The answer is 7,088.

<u>Example 2</u>

Solve the problem 1,762 + 3,519.

Step 1: Split the problem into two parts by drawing or imagining a line down the middle.

$$\begin{array}{r|l} 17 & 62 \\ +35 & 19 \\ \hline & \end{array}$$

Step 2: Add the numbers on the left side of the split: 17 + 35 = 52.

$$\begin{array}{r|l} 17 & 62 \\ +35 & 19 \\ \hline 52 & \end{array}$$

Step 3: Add the numbers on the right side of the split: 62 + 19 = 81. Bring the two parts together.

$$\begin{array}{r|l} 17 & 62 \\ +35 & 19 \\ \hline 52 & 81 \end{array}$$

Solution: The answer is 5,281.

<u>Example 3</u>

Solve the problem 5,235 + 8,997.

Step 1: Split the problem into two parts by drawing or imagining a line down the middle.

$$\begin{array}{r|l} 52 & 35 \\ +89 & 97 \\ \hline & \end{array}$$

Step 2: Add the numbers on the left side of the split: 52 + 89 = 141.

$$\begin{array}{r|l} 52 & 35 \\ +89 & 97 \\ \hline 141 & \end{array}$$

Step 3: Add the numbers on the right side of the split: 35 + 97 = 132. Write or think of the 1 as smaller; this is the carryover.

$$\begin{array}{r|l} 52 & 35 \\ +89 & 97 \\ \hline 141 & {}_{1}32 \end{array}$$

Step 4: Carry over the 1 from the right to the left side of the split. Bring the two parts together.

$$\begin{array}{r|l} 52 & 35 \\ +89 & 97 \\ \hline 141 & {}_{1}32 \end{array}$$

$$14232$$

Solution: The answer is 14,232.

QUICK TIP

As you gain confidence applying number splitting, try doing the problems in your head rather than on a piece of paper. It's easier than you think!

The Least You Need to Know

- To find quick answers for problems with two- and three-digit numbers, you can add the columns from left to right and then combine the middle digits.
- Changing numbers to 10 or a multiple of 10 in a problem can help you get a sum much faster. You then simply subtract from the sum how much was needed to get to 10 or the multiple.
- You can break a problem down into more manageable parts through number splitting.

CHAPTER

3

Subtraction

In This Chapter

- Subtracting left to right
- Subtracting from powers of 10
- Learn subtraction by the all from 9 and last from 10 method

Generally speaking, subtraction is considered to be difficult because of the carryovers and the concept of borrowing. Even though you deal with subtraction early on in your student life, you may still have a kind of mental aversion to it as you grow up.

In this chapter, I show you some shortcuts to get answers to subtraction problems. Soon, you won't mind subtracting at all!

Left-to-Right Subtraction

Unlike the traditional system of subtraction, which goes from right to left, subtracting from left to right helps you deal with carryovers in a faster and easier fashion.

Example 1

Solve the problem 724 – 261.

Step 1: Subtract the first column: 7 – 2 = 5.

```
 724
-261
----
 5
```

Step 2: Subtract the second column. Because 2 is less than 6, you reduce the answer in the first column by 1, changing 5 to 4. You then place that carryover 1 with the second column, which changes 2 to 12, and subtract: 12 – 6 = 6.

```
 7¹24
- 261
-----
 5̸46
```

Step 3: Subtract the third column: 4 – 1 = 3.

```
 7¹24
- 261
-----
 5̸463
```

Solution: The answer is 463.

QUICK TIP

In subtraction, one common problem you'll run into is the top number being smaller than the bottom number. In such cases, you go back one step, reduce the number in the previous column by 1, and carry over the 1 to the current step.

<u>Example 2</u>

Solve the problem 647 – 294.

Step 1: Subtract the first column: 6 – 2 = 4.

$$\begin{array}{r} 647 \\ -\,294 \\ \hline 4 \end{array}$$

Step 2: Subtract the second column. Because 4 is less than 9, you reduce the answer in the first column by 1, changing 4 to 3. You then place that carryover 1 with the second column, which changes 4 to 14, and subtract: 14 – 9 = 5.

$$\begin{array}{r} 6^{1}47 \\ -\,294 \\ \hline \cancel{4}35 \end{array}$$

Step 3: Subtract the third column: 7 – 4 = 3.

$$\begin{array}{r} 6^{1}47 \\ -\,294 \\ \hline \cancel{4}353 \end{array}$$

Solution: The answer is 353.

<u>Example 3</u>

Solve the problem 4,247 – 1,763.

Step 1: Subtract the first column: 4 – 1 = 3.

$$\begin{array}{r} 4247 \\ -\,1763 \\ \hline 3 \end{array}$$

Step 2: Subtract the second column. Because 2 is less than 7, you reduce the answer in the first column by 1, changing 3 to 2. You then place that carryover 1 with the second column, which changes 2 to 12, and subtract: 12 – 7 = 5.

$$\begin{array}{r} 4^{1}247 \\ -\,1763 \\ \hline \cancel{3}25 \end{array}$$

Step 3: Subtract the third column. Because 4 is less than 6, you reduce the answer in the second column by 1, changing 5 to 4. You then place that carryover 1 with the third column, which changes 4 to 14, and subtract: 14 – 6 = 8.

$$\begin{array}{r} 4^{1}2^{1}47 \\ -\,1763 \\ \hline \cancel{3}2\cancel{5}48 \end{array}$$

Step 4: Subtract the fourth column: 7 – 3 = 4.

$$\begin{array}{r} 4^{1}2^{1}47 \\ -\,1763 \\ \hline \cancel{3}2\cancel{5}484 \end{array}$$

Solution: The answer is 2,484.

Subtraction with Numbers Near 10 or a Multiple of 10

Numbers near a base or multiple of 10—such as 9, 19, 28, 38, 47, and so on—are very quick and easy to subtract. All you have to do is round the number up to the multiple of 10 it's closest to and solve; after that, you add back in the difference.

<u>Example 1</u>

Solve the problem 52 – 8.

Step 1: Change 8 to 10 and subtract.

52 – 10 = 42

Step 2: Because 8 is 2 below 10, add 2 back to get the answer.

42 + 4 = 44

Solution: The answer is 44.

QUICK TIP

Don't forget to add back in the difference between the multiple of 10 and the number you replaced it with.

Example 2

Solve the problem 82 – 49.

Step 1: Change 49 to 50 and subtract.

82 – 50 = 32

Step 2: Because 49 is 1 below 50, add 1 back to get the answer.

32 + 1 = 33

Solution: The answer is 33.

Example 3

Solve the problem 162 – 77.

Step 1: Change 77 to 80 and subtract.

162 – 80 = 82

Step 2: Because 77 is 3 below 80, add 3 back to get the answer.

82 + 3 = 85

Solution: The answer is 85.

Number Splitting

As you learned in Chapter 2, number splitting can help you split a problem in more manageable parts. You can employ this method when subtracting large numbers.

The following examples show you how you can use number splitting to solve subtraction problems with larger numbers.

Example 1

Solve the problem 6,389 – 4,245.

Step 1: Split the problem into two parts by drawing or imagining a line down the middle.

$$\begin{array}{r|l} 63 & 89 \\ -\ 42 & 45 \\ \hline & \end{array}$$

Step 2: Subtract the numbers on the left side of the split: 63 – 42 = 21.

$$\begin{array}{r|l} 63 & 89 \\ -\ 42 & 45 \\ \hline 21 & 44 \end{array}$$

Step 3: Subtract the numbers on the right side of the split: 89 – 45 = 44. Bring the two parts together.

$$\begin{array}{r|l} 63 & 89 \\ -\ 42 & 45 \\ \hline 21 & 44 \end{array}$$

Solution: The answer is 2,144.

Example 2

Solve the problem 6,889 – 1,836.

Step 1: Split the problem into two parts by drawing or imagining a line down the middle.

$$\begin{array}{r|l} 68 & 89 \\ -\ 18 & 36 \\ \hline & \end{array}$$

Step 2: Subtract the numbers on the left side of the split: 68 – 18 = 50.

68	89
- 18	36
50	

Step 3: Subtract the numbers on the right side of the split: 89 – 36 = 53. Bring the two parts together.

68	89
- 18	36
50	53

Solution: The answer is 5,053.

Example 3

Solve the problem 4,468 – 2,286.

Step 1: Split the problem. Here, note that if you split this problem from the middle, you'd have to do 68 – 86, which would give you a negative answer. Therefore, you should split the numbers into more pieces to avoid any negative numbers.

4	46	8
- 2	28	6

Step 2: Subtract the left portion: 4 – 2 = 2.

4	46	8
- 2	28	6
		2

Step 3: Subtract the middle portion: 46 – 28 = 18.

4	46	8
- 2	28	6
	21	8

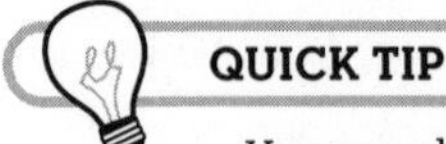

QUICK TIP

You can always use what you learned earlier about rounding up to 10 or a multiple of 10 to help you solve the pieces of the number splits. For example, with 46 - 28, you can change the 28 to 30 to solve 46 - 30 = 16. You then add in 2 to get the answer: 16 + 2 = 18.

Step 4: Subtract the right portion: 8 – 6 = 2. Bring the parts together.

4	46	8
- 2	28	6
2	18	2

Solution: The answer is 2,182.

Subtraction with the All from 9 and Last from 10 Method

For subtraction, I'd like to introduce you to a method called *all from 9 and last from 10.* This method can be used when the digit on the top row is less than the digit below it. The following examples show how this method is applied in different scenarios.

Subtracting from a Power of 10

To subtract from a power of 10 using the all from 9 and last from 10 method, going from right to left, you subtract every digit except the last from 9. The last digit is then subtracted from 10.

Example 1

Solve the problem 1,000 – 283.

Step 1: Subtract the 2 in the first column from 9.

9 - 2 = 7

Step 2: Subtract the 8 in the second column from 9.

9 - 8 = 1

Step 3: Because the third column is the last column in this number, subtract 3 from 10.

10 – 3 = 7

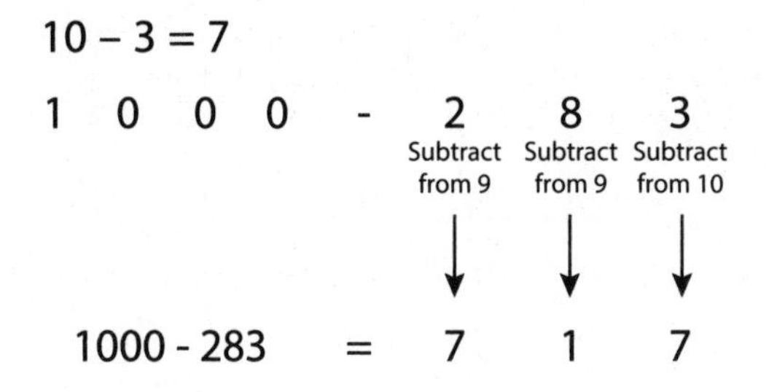

Solution: The answer is 717.

<u>Example 2</u>

Solve the problem 1,000 – 476.

Step 1: Subtract the 4 in the first column from 9.

9 – 4 = 5

Step 2: Subtract the 7 in the second column from 9.

9 – 7 = 2

Step 3: Because the third column is the last column in this number, subtract 6 from 10.

10 – 6 = 4

1 0 0 0 - 4 7 6

Subtract from 9 ↓ Subtract from 9 ↓ Subtract from 10 ↓

1000 - 476 = 5 2 4

Solution: The answer is 524.

Example 3

Solve the problem 10,000 – 2,783.

Step 1: Subtract the 2 in the first column from 9:

9 – 2 = 7

Step 2: Subtract the 7 in the second column from 9.

9 – 7 = 2

Step 3: Subtract the 8 in the third column from 9.

9 – 8 = 1

Step 4: Because the third column is the last column in this number, subtract 3 from 10.

10 – 3 = 7

1 0 0 0 0	-	2	7	8	3
		Subtract from 9	Subtract from 9	Subtract from 9	Subtract from 10
		↓	↓	↓	↓
10000 - 2783	=	7	2	1	7

Solution: The answer is 7,217.

Subtracting a Number Ending in Zero from a Power of 10

Subtracting from a power of 10 when your other number has a zero at the end is a bit different. In this case, you subtract every digit except the second to last one from 9; the second to last is then subtracted from 10. The 0 is then simply included at the end of the number.

Example 1

Solve the problem 1,000 – 280.

Step 1: Subtract the 2 in the first column from 9.

9 – 2 = 7

Step 2: Subtract the 8 in the second column from 10.

10 – 8 = 2

Step 3: Include the 0 from the third column on the end of the answer; this changes 72 to 720.

1 0 0 0	-	2	8	0
		Subtract from 9	Subtract from 10	
		↓	↓	↓
1000 - 280	=	7	2	0

Solution: The answer is 720.

<u>Example 2</u>

Solve the problem 1,000 – 370.

Step 1: Subtract the 3 in the first column from 9.

$9 - 3 = 6$

Step 2: Subtract the 7 in the second column from 10.

$10 - 7 = 3$

Step 3: Include the 0 from the third column on the end of the answer; this changes 63 to 630.

1 0 0 0	-	3	7	0
		Subtract from 9	Subtract from 10	
		↓	↓	↓
1000 - 370	=	6	3	0

Solution: The answer is 630.

<u>Example 3</u>

Solve the problem 10,000 – 5,270.

Step 1: Subtract the 5 in the first column from 9.

$9 - 5 = 4$

Step 2: Subtract the 2 in the second column from 9.

$9 - 2 = 7$

Step 3: Subtract the 7 in the third column from 10.

$10 - 7 = 3$

Step 4: Include the 0 from the fourth column on the end of the answer; this changes 473 to 4,730.

10000	-	5	2	7	0
		Subtract from 9	Subtract from 9	Subtract from 10	
		↓	↓	↓	↓
10000 - 5270 =		4	7	3	0

Solution: The answer is 4,730.

Subtracting When Neither Number is a Power of 10

For a problem setup not involving a power of 10, you have to take some extra steps with this method to get your answer. Also, unlike what you've learned for other methods in this chapter, you have to work the problems from right to left.

<u>Example 1</u>

Solve the problem 651 – 297.

Step 1: Subtract the ones column. You can see that 1 is less than 7. To avoid a carryover, you first do the subtraction in reverse: 7 – 1 = 6. You then subtract the sum from 10: 10 – 6 = 4.

$$\begin{array}{r} 651 \\ -297 \\ \hline 4 \end{array}$$

QUICK TIP

Let me repeat the process for the first step in the first example, because it takes a little time to sink in. You start with the right column. In this column, 7 is more than 1. Instead of borrowing, subtract in reverse: 7 - 1. This gives you 6.

You now apply the all from 9 and last from 10 method. Because the number in the rightmost column is the last digit, you subtract 10 from the sum to get your answer: 10 - 6. This gives you 4, which is what you write in the ones column.

Step 2: Subtract the tens column. The 5 is less than 9, so start by subtracting in reverse: 9 – 5 = 4. You then subtract the sum from 9: 9 – 4 = 5.

$$\begin{array}{r} 651 \\ -\,297 \\ \hline 54 \end{array}$$

Step 3: Subtract the hundreds column. The 6 is greater than 2, so subtract as you would normally: 6 – 2 = 4. Because you've done so many adjustments to the previous digits, you need to subtract 1 from the sum: 4 – 1 = 3.

$$\begin{array}{r} 651 \\ -\,297 \\ \hline 354 \end{array}$$

Solution: The answer is 354.

<u>Example 2</u>

Solve the problem 425 – 168.

Step 1: Subtract the ones column. Because 5 is less than 8, subtract in reverse: 8 – 5 = 3. You then subtract the sum from 10: 10 – 3 = 7.

$$\begin{array}{r} 425 \\ -\,168 \\ \hline 7 \end{array}$$

Step 2: Subtract the tens column. The 2 is less than 6, so first subtract in reverse: 6 – 2 = 4. You then need to subtract 9 from the sum: 9 – 4 = 5.

$$\begin{array}{r} 425 \\ -\,168 \\ \hline 57 \end{array}$$

Step 3: Subtract the hundreds column. The 4 is more than 1, so you can subtract normally: 4 – 1 = 3. Because you've done so many adjustments to the previous digits, you need to subtract 1 from the sum: 3 – 1 = 2.

$$\begin{array}{r} 425 \\ -\,168 \\ \hline 257 \end{array}$$

Solution: The answer is 257.

<u>Example 3</u>

Solve the problem 7,643 – 4,869.

Step 1: Subtract the ones column. The 3 is less than 9, so start by subtracting in reverse: 9 – 3 = 6. You next subtract the sum from 10: 10 – 6 = 4.

$$\begin{array}{r} 7643 \\ -\,4869 \\ \hline 4 \end{array}$$

Step 2: Subtract the tens column. Because 4 is less than 6, subtract in reverse: 6 – 4 = 2. Now subtract the sum from 9: 9 – 2 = 7.

$$\begin{array}{r} 7643 \\ -\,4869 \\ \hline 74 \end{array}$$

Step 3: Subtract the hundreds column. The 6 is more than 8, so first subtract in reverse: 8 – 6 = 2. You then need to subtract the sum from 9: 9 – 2 = 7.

$$\begin{array}{r} 7643 \\ -\,4869 \\ \hline 774 \end{array}$$

Step 4: Subtract the thousands column. The 7 is more than 4, you can subtract normally: 7 – 4 = 3. Because of the adjustments to the previous digits, you then subtract 1 from the sum: 3 – 1 = 2.

$$\begin{array}{r} 7643 \\ -\,4869 \\ \hline 2774 \end{array}$$

Solution: The answer is 2,774.

Starting the Method Later in the Problem

Sometimes, you may need to apply the all from 9 and last from 10 method starting in the middle of the sum. This happens when the top number is larger than the bottom in at least the first column of the problem.

Example 1

Solve the problem 7,251 – 2,790.

Step 1: Subtract the ones column. The 1 on top is more than the 0 at the bottom, so you subtract normally: 1 – 0 = 1.

$$\begin{array}{r} 7251 \\ -\,2790 \\ \hline 1 \end{array}$$

Step 2: Subtract the tens column. Because 5 is greater than 9, subtract in reverse: 9 – 5 = 4. You didn't have to apply the method to the previous digit, so you subtract this sum from 10: 10 – 4 = 6.

$$\begin{array}{r} 7251 \\ -\,2790 \\ \hline 61 \end{array}$$

Step 3: Subtract the hundreds column. The 2 is less than 7, so subtract in reverse: 7 – 2 = 5. You then subtract the sum from 9: 9 – 5 = 4.

$$\begin{array}{r} 7251 \\ -\,2790 \\ \hline 461 \end{array}$$

Step 4: Subtract the thousands column. The 7 is greater than 2, so you can subtract normally: 7 – 2 = 5. Because of all the adjustments you made, you then have to subtract 1 from the sum: 5 – 1 = 4.

$$\begin{array}{r} 7251 \\ -\,2790 \\ \hline 4461 \end{array}$$

Solution: The answer is 4,461.

Example 2

Solve the problem 92,138 – 27,804.

Step 1: In the ones and tens columns, the digits in the top row are greater than those below them. Therefore, you can subtract them normally: 8 – 4 = 4 and 3 – 0 = 3.

$$\begin{array}{r} 92138 \\ -\,27804 \\ \hline 34 \end{array}$$

Step 2: Subtract the hundreds column. Because 1 is less than 8, subtract in reverse: 8 – 1 = 7. You didn't have to apply the method to the previous two digits, so subtract the sum from 10: 10 – 7 = 3.

$$\begin{array}{r} 92138 \\ -\,27804 \\ \hline 334 \end{array}$$

Step 3: Subtract the thousands column. The 2 is less than 7, so subtract in reverse: 7 – 2 = 5. You then subtract the sum from 9: 9 – 5 = 4.

$$\begin{array}{r} 92138 \\ -\,27804 \\ \hline 4334 \end{array}$$

Step 4: Subtract the ten thousands column. Because 9 is greater than 2, you can subtract normally: 9 – 2 = 7. You then have to subtract 1 from the sum due the earlier adjustments: 7 – 1 = 6.

$$\begin{array}{r} 92138 \\ -\ 27804 \\ \hline 64334 \end{array}$$

Solution: The answer is 64,334.

Ending the Method Earlier in the Problem

When at least the last column of the problem has a top number that's greater than the bottom number, you have to finish the method earlier in the problem.

SPEED BUMP

It pays to inspect the problem first before you begin the all from 9 and last from 10 method. You don't want to automatically start using it, only to realize you began or ended too early. However, if you trip up, just try again—after all, practice makes perfect!

Example 1

Solve the problem 638,475 – 429,763.

Step 1: Subtract the ones and tens columns. In the ones column, because 5 is greater than 3, you can subtract normally: 5 – 3 = 2. The 7 is greater than 6 in the tens column, so you can also subtract this normally: 7 – 6 = 1.

$$\begin{array}{r} 638475 \\ -\ 429763 \\ \hline 12 \end{array}$$

Step 2: Subtract the hundreds column. The 4 is less than 7, so subtract in reverse: 7 – 4 = 3. You didn't have to apply the method to the previous two digits; therefore, you subtract the sum from 10: 10 – 3 = 7.

```
 638475
-429763
-------
    712
```

Step 3: Subtract the thousands column. Because 8 is less than 9, you need to subtract in reverse: 9 – 8 = 1. You then subtract the sum from 9: 9 – 1 = 8.

```
 638475
-429763
-------
   8712
```

Step 4: Subtract the ten thousands column. The 3 is more than 2, so you can simply subtract: 3 – 2 = 1. Because the numbers in the next column can subtract normally, this is where you finish applying the method. You subtract 1 to account for the adjustments: 1 – 1 = 0.

```
 638475
-429763
-------
  08712
```

Step 5: Subtract the hundred thousands column. The 6 on top is greater than the 4, so subtract normally: 6 – 4 = 2.

```
 638475
-429763
-------
 208712
```

Solution: The answer is 208,712.

Example 2

Solve the problem 4,826,495 – 3,717,871.

Step 1: Subtract the ones and tens columns. In the ones column, the 5 is greater than 1, so you can subtract normally: 5 – 1 = 4. The 9 on top is more than 7 in the tens column, so you can subtract again as usual: 9 – 7 = 2.

$$\begin{array}{r} 4826495 \\ -\,3717871 \\ \hline 24 \end{array}$$

Step 2: Subtract the hundreds column. Because 4 is less than 8, you first subtract in reverse: 8 – 4 = 4. You didn't apply the method to the previous digits, so you subtract the sum from 10: 10 – 4 = 6.

$$\begin{array}{r} 4826495 \\ -\,3717871 \\ \hline 624 \end{array}$$

Step 3: Subtract the thousands column. The 6 is less than 7, so subtract in reverse: 7 – 6 = 1. You then subtract the sum from 9: 9 – 1 = 8.

$$\begin{array}{r} 4826495 \\ -\,3717871 \\ \hline 8624 \end{array}$$

Step 4: Subtract the ten thousands column. The 2 on top is more than the 1 below it, so subtract normally: 2 – 1 = 1. Because the top digit is larger than the bottom digit, this is where you stop applying the method in the problem. You subtract 1 to account for the earlier adjustments: 1 – 1 = 0.

$$\begin{array}{r} 4826495 \\ -\,3717871 \\ \hline 08624 \end{array}$$

Step 5: Subtract the hundred thousands and millions columns. Both of them have top digits larger than the bottom digits, so subtract as normal: 8 – 7 = 1 and 4 – 3 = 1.

$$
\begin{array}{r}
4826495 \\
-\ 3717871 \\
\hline
1108624
\end{array}
$$

Solution: The answer is 1,108,624.

The Least You Need to Know

- Left-to-right subtraction can help you deal with carry-overs in a quick and simple manner.
- It's much easier to round a number to 10 or a multiple of 10 before subtracting and then add in the difference than simply trying to subtract as is.
- The all from 9 and last from 10 method is applied when the top number is less than the bottom number in a problem.

CHAPTER

4

Division

In This Chapter

- The flag method of division
- Using altered remainders to deal with negative numbers
- Finding the decimal value of fractions with auxiliary fractions

Division can be a trying process without a calculator, especially when you don't have single-digit divisors. But with a couple new processes in your arsenal, you will soon be doing division faster than ever. And who knows? You may even learn to like it!

In this chapter, I show you how the flag method, altered remainders, and auxiliary fractions can speed up division for you.

Division with the Flag Method

To help you with division beyond two one-digit numbers, I'd like to introduce you to the *flag method*. To begin the flag method, you write the divisor down in a "flag" formation—for example, if it's a two-digit number, the first digit is the "flag pole," and the second digit is the "flag." The following shows how it looks using 31 as a divisor.

1 → Flag
3
→ Flag Pole

You then follow two basic rules to get the answer:

1. Divide by the flag pole.
2. Subtract by the flag times the previous quotient digit; think of it like Digit – (Flag × Previous Quotient Digit).

So there are only two basic rules for applying this method: divide and subtract. It's a cyclical process, so if you divide in a step, you subtract in the next step, and so on.

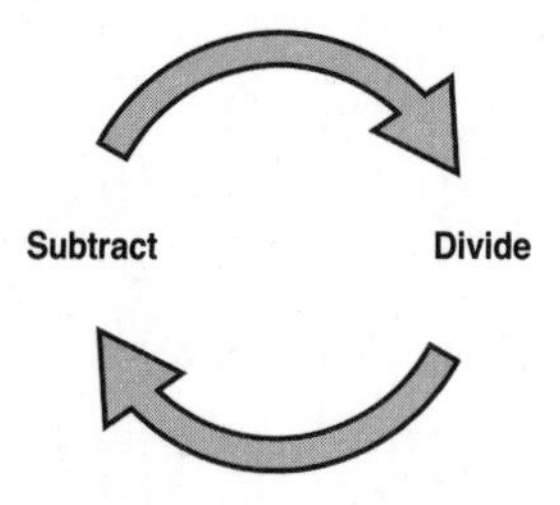

Just keep this cycle in mind, and you'll have it down in no time!

Two-Digit Divisors

If you have a two-digit divisor, you set up the problem with a line between the divisor and dividend. You then put another line in to indicate where the decimal point goes in the answer. The decimal point is based on how many digits you have in the flag; for a two-digit number, you only have one digit in the flag, so you count over one from the left in the divisor and place the line. You work the problem by following the two simple steps of the flag method: divide and subtract.

Example 1

Solve the problem 848 ÷ 31 to one decimal point.

Step 1: Lay out the problem. Because 848 is the dividend, put a line before it. For the divisor 31, put it to the left of the line, with the 3 serving as the flag pole and the 1 serving as the flag. You only have one digit in the flag, so put a line one to the left of the dividend to indicate where the decimal point goes.

$$\begin{array}{l|l|l} 3^1 & 8\ 4 & 8 \\ \hline & & \end{array}$$

Step 2: Divide 8 by the flag pole, 3. This gives you a quotient digit of 2 with a remainder of 2 (because 3 × 2 = 6, which is the highest multiple of 3 that can go into 8). Put down 2 as and prefix the remainder to 4 so it becomes 24.

$$\begin{array}{l|l|l} 3^1 & 8_2 4 & 8 \\ \hline & 2 & \end{array}$$

Step 3: Subtract 24 by the flag times the previous quotient digit.

$$24 - (1 \times 2) = 22$$

Step 4: Divide 22 by the flag pole: 22 ÷ 3 = 7, remainder 1. Put down 7 and prefix the remainder to 8 so it becomes 18.

$$\begin{array}{l|l|l} 3^1 & 8_2 4 & {}_1 8 \\ \hline & 2\ 7 & \end{array}$$

Step 5: Subtract 18 by the flag times the previous quotient digit.

$$18 - (1 \times 7) = 11$$

Step 6: Divide 11 by the flag pole: 11 ÷ 3 = 3, remainder 2. Put down 3 and carry over the 2.

$$\begin{array}{l|l|l} 3^1 & 8_2 4 & {}_1 8_2 \\ \hline & 2\ 7 & 3 \end{array}$$

Solution: The answer is 27.3.

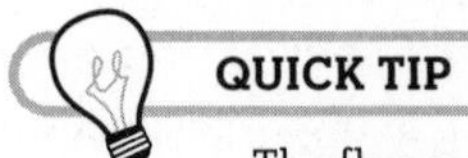

The flag method can be a little hard to pick up at first, but if you devote 15 to 20 minutes of your time to practicing it, you should get it. I must confess, when I first looked at this method, I was totally lost! After 10 minutes, though, I was able to understand and apply it. Just remember the two rules: divide and subtract.

Example 2

Solve the problem 5,576 ÷ 25 to two decimal places.

Step 1: Lay out the problem. Because 5,576 is the dividend, put a line before it. For the divisor 25, put it to the left of the line, with the 2 serving as the flag pole and the 5 serving as the flag. You only have one digit in the flag, so put a line one to the left of the dividend to indicate where the decimal point goes.

2^5	5 5 7	6

Step 2: Divide 5 by the flag pole, which is 2. This gives you 2 as the quotient and 1 as the remainder (because 2 × 2 = 4, which is the highest multiple of 2 that can go into 8). Put down 2 and prefix the remainder of 1 to 5 so it becomes 15.

2^5	$5\ {}_{1}5\ 7$	6
	2	

Step 3: Subtract 15 by the flag times the previous quotient digit.

15 – (5 × 2) = 5

Step 4: Divide 5 by the flag pole: 5 ÷ 2 = 2, remainder 1. Put down 2 and prefix the remainder of 1 to 7 so it becomes 17.

2^5	$5{}_{1}5{}_{1}7$	6
	2 2	

Step 5: Subtract by the flag times the previous quotient digit.

17 – (5 × 2) = 7

Step 6: Divide 7 by the flag pole: 7 ÷ 3, remainder 1. Put 3 down as the quotient and prefix the remainder of 1 to 6 so it becomes 16.

2^5	5 $_1$5 $_1$7	$_1$6
	2 2 3.	

Step 7: Subtract 16 by the flag times the previous quotient digit.

$16 - (5 \times 3) = 1$

Step 8: Divide 1 by the flag pole: 1 ÷ 2 = 0, remainder 1. Put down 0 and add a 0 above so you can prefix the 1 to it, making it 10.

2^5	5 $_1$5 $_1$7	$_1$6 $_1$0
	2 2 3.	0

Step 9: Subtract 10 by the flag times the previous quotient digit.

$10 - (5 \times 0) = 10$

Step 10: Divide 10 by the flag pole: 10 ÷ 2 = 5. Put down 5.

2^5	5 $_1$5 $_1$7	$_1$6 $_1$0
	2 2 3.	0 5

Solution: The answer is 223.05.

<u>Example 3</u>

Solve the problem 2,924 ÷ 72 to two decimal places.

Step 1: Lay out the problem. Because 2,924 is the dividend, put a line before it. For the divisor 72, put it to the left of the line, with the 7 serving as the flag pole and the 2 serving as the flag. You only have one digit in the flag, so put a line one to the left of the dividend to indicate where the decimal point goes.

7^2	2 9 2 4

Step 2: Divide 29 by the flag pole, 7. This gives you 4 as a quotient digit and a remainder of 1 (because 7 × 4 = 28, which is the highest multiple of 7 that can go into 29). Put down 4 and prefix the remainder of 1 to 2 so it becomes 12.

$$\begin{array}{c|cccc} 7^2 & 2 & 9 & {}_{1}2 & 4 \\ \hline & 4 & & & \end{array}$$

Step 3: Subtract 12 by the flag times the previous quotient digit.

$$12 - (2 \times 4) = 12 - 8 = 4$$

Step 4: Divide 4 by the flag pole: 4 ÷ 7 = 0, remainder 4. Put down 0 and prefix the remainder of 4 to the 4 above so it becomes 44.

$$\begin{array}{c|ccc|c} 7^2 & 2 & 9 & {}_{1}2 & {}_{4}4 \\ \hline & 4 & 0. & & \end{array}$$

Step 5: Subtract 44 by the flag times the previous quotient digit.

$$44 - (2 \times 0) = 44$$

Step 6: Divide 44 by the flag pole: 44 ÷ 7 = 6, remainder 2. Put down 6 and carry over the remainder of 2.

$$\begin{array}{c|ccc|c} 7^2 & 2 & 9 & {}_{1}2 & {}_{4}4\ {}_{2} \\ \hline & 4 & 0. & & 6 \end{array}$$

Step 7: To find out the answer to two decimal places, put 0 after the remainder of 2 and subtract 20 by the flag times the previous quotient digit.

$$20 - (2 \times 6) = 8$$

Step 8: Divide 8 by the flag pole: 8 ÷ 7 = 1 and 1 remainder.

$$\begin{array}{c|ccc|cc} 7^2 & 2 & 9 & {}_{1}2 & {}_{4}4 & {}_{2}0_{1} \\ \hline & 4 & 0. & & 6 & 1 \end{array}$$

Solution: The answer is 40.61.

Three-Digit Divisors

Now that you've gotten more comfortable with the flag method, let's step up to three-digit divisors. In this case, you'll have one digit in the flag pole and two digits in the flag. This also means the decimal point is two digits from the left of the dividend. You'll continue the division and subtraction cycle, but the way the digits in the flag are used is slightly different.

When you first subtract, it's only from the *first digit* of the flag times the previous quotient digit. However, for subsequent subtraction portions, you do the following:

> Digit – [(First Digit of Flag × Previous Quotient Digit) + (Second Digit of Flag × Quotient Digit Before Previous Quotient Digit)]

This probably looks pretty confusing, so let me walk you through some examples so you can fully understand the process.

<u>Example 1</u>

Solve the problem 888 ÷ 672 to two decimal places.

Step 1: Lay out the problem. Because 888 is the dividend, put a line before it. For the divisor 672, put it to the left of the line, with the 6 serving as the flag pole and the 72 serving as the flag. You have two digits in the flag, so put a line two to the left of the dividend to indicate where the decimal point goes.

6^{72}	8	8 8

Step 2: Divide 8 by the flag pole, which is 6. This gives you a quotient digit of 1 and a remainder of 2 (because 6 × 1 = 6, which is the highest multiple of 6 that can go into 8). Put down 1 and prefix the remainder of 2 to the 8 so it becomes 28.

6^{72}	8	$_{2}8$ 8
	1	

Step 3: Subtract 28 by the *first digit* of the flag times the previous quotient digit.

$$28 - (1 \times 7) = 28 - 7 = 21$$

Step 4: Divide 21 by the flag pole: 21 ÷ 6 = 3, remainder 3. Put down 3 and prefix the remainder of 3 to 8 so it becomes 38.

6^{72}	8	$_2 8 _3 8$
	1	3

Step 5: Subtract 38 by the first digit of the flag times the previous quotient plus the second digit of the flag times the quotient digit before that. In this case, you have (7 × 3) and (2 × 1) in the brackets.

$$38 - [(3 \times 7) + (1 \times 2)] = 38 - 23 = 15$$

Step 6: Divide 15 by the flag pole: 15 ÷ 6 = 2, remainder 3. Put down 2 and put the remainder of 3 after the 8.

6^{72}	8	$_2 8 _3 8 _3$
	1.	32

Solution: The answer is 1.32.

<u>Example 2</u>

Solve the problem 70,319 ÷ 823 to two decimal places.

Step 1: Lay out the problem. Because 70,319 is the dividend, put a line before it. For the divisor 823, put it to the left of the line, with the 8 serving as the flag pole and the 23 serving as the flag. You have two digits in the flag, so put a line two to the left of the dividend to indicate where the decimal point goes.

8^{23}	703	19

Step 2: Divide 70 by the flag pole, 8. This gives you 8 as the quotient digit with a remainder of 6 (because 8 × 8 = 64, which is the highest multiple of 8 that can go into 70). Put down 8 and prefix the remainder of 6 to 3 so it becomes 63.

8^{23}	$70_{6}3$	19
	8	

Step 3: Subtract 63 by the *first digit* of the flag times the previous quotient digit.

$$63 - (2 \times 8) = 63 - 16 = 47$$

SPEED BUMP

In the third step, do *not* subtract 63 - (**23** × 8). Use only the first digit of the flag in the multiplication portion: 63 - (**2** × 8).

Step 4: Divide 47 by the flag pole: 47 ÷ 8 = 5, remainder 7. Put down 5 and prefix the remainder of 7 to 1 so it becomes 71.

8^{23}	$70_{6}3$	${}_{7}19$
	8 5	

Step 5: Subtract 71 by the first digit of the flag times the previous quotient plus the second digit of the flag times the quotient digit before that. For this problem, this means you have (2 × 5) and (3 × 8) in the brackets.

$$71 - [(2 \times 5) + (3 \times 8)] = 71 - 34 = 37$$

Step 6: Divide 37 by the flag pole: 37 ÷ 8 = 4, remainder 5. Put down 4 and prefix the remainder of 5 to 9 so it becomes 59.

8^{23}	$70_{6}3$	$_{7}1_{5}9$
	8 5.	4

Step 7: Subtract 59 by the first digit of the flag times the previous quotient plus the second digit of the flag times the quotient digit before that. Here, you put (2 × 4) and (3 × 5) in the brackets.

$$59 - [(4 \times 2) + (5 \times 3)] = 59 - 23 = 36$$

Step 8: Divide 36 by the flag pole: 36 ÷ 8 = 4, remainder 4. Put down 4 and put the remainder of 4 after the 9.

8^{23}	$70_{6}3$	$_{7}1_{5}9$
	8 5.	4 4

Solution: The answer is 85.44.

<u>Example 3</u>

Solve the problem 10,643 ÷ 743 to two decimal places.

Step 1: Lay out the problem. Because 10,643 is the dividend, put a line before it. For the divisor 743, put it to the left of the line, with the 7 serving as the flag pole and the 43 serving as the flag. You have two digits in the flag, so put a line two to the left of the dividend to indicate where the decimal point goes.

7^{43}	1 0 6	43

Step 2: Divide 10 by the flag pole, 7. This gives you a quotient of 1 and a remainder of 3 (because 7 × 1 = 7, which is the highest multiple of 7 that can go into 10). Put down 1 and prefix the remainder of 3 to 6 so it becomes 36.

7^{43}	$1\ 0\ {}_{3}6$	43
	1	

Step 3: Subtract 36 by the *first digit* of the flag times the previous quotient digit.

$$36 - (4 \times 1) = 32$$

Step 4: Divide 32 by the flag pole: $32 \div 7 = 4$, remainder 4. Put down 4 and prefix the remainder of 4 to the 4 above so it becomes 44.

7^{43}	$1\ 0\ {}_{3}6$	${}_{4}43$
	1 4	

Step 5: Subtract 44 by the first digit of the flag times the previous quotient plus the second digit of the flag times the quotient digit before that. In this case, you have (4×4) and (3×1) in the brackets.

$$44 - [(4 \times 4) + (3 \times 1)] = 44 - 19 = 25$$

Step 6: Divide 25 by the flag pole: $25 \div 7 = 3$, remainder 4. Put down 3 and prefix the remainder of 4 to 3 so it becomes 43.

7^{43}	$1\ 0\ {}_{3}6$	${}_{4}4{}_{4}3$
	1 4.	3

Step 7: Subtract 43 by the first digit of the flag times the previous quotient plus the second digit of the flag times the quotient digit before that. In this case, you have (4×3) and (3×4) in the brackets.

$$43 - [(4 \times 3) + (3 \times 4)] = 43 - 24 = 19$$

Step 8: Divide 19 by the flag pole: 19 ÷ 7 = 2, remainder 5. Put down 2 and put the remainder of 5 after the 3.

7^{43}	$1\,0\,{}_{3}6$	${}_{4}4{}_{4}3{}_{5}$
	1 4.	32

Solution: The answer is 14.32.

Four-Digit Divisors

The flag method for a four-digit divisor is very similar to what you do for the three-digit divisor. The main difference is that you now have two digits in the flag pole, and you use the two digits together to divide the numbers. The following examples show you what you need to do with division problems involving four-digit divisors.

Example 1

Solve the problem 4,213 ÷ 1,234 to two decimal places.

Step 1: Lay out the problem. Because 4,213 is the dividend, put a line before it. For the divisor 1,234, put it to the left of the line, with the 12 serving as the flag pole and the 34 serving as the flag. You have two digits in the flag, so put a line two to the left of the dividend to indicate where the decimal point goes.

12^{34}	42	13

Step 2: Divide 42 by the flag pole, 12. This gives you a quotient of 3 and a remainder of 6 (because 12 × 3 = 36, which is the highest multiple of 12 that can go into 42). Put down 3 and prefix the remainder of 6 to 1 so it becomes 61.

12^{34}	42	${}_{6}13$
	3	

Step 3: Subtract 61 by the first digit of the flag times the previous quotient digit.

$$61 - (3 \times 3) = 61 - 9 = 52$$

Step 4: Divide 52 by the flag pole: 52 ÷ 12 = 4, remainder 4. Put down 4 and prefix the remainder of 4 to 3 so it becomes 43.

12^{34}	42	${}_{6}1_{4}3$
	3.	4

Step 5: Subtract 43 by the first digit of the flag times the previous quotient plus the second digit of the flag times the quotient digit before that. In this case, you have (3 × 4) and (4 × 3) in the brackets.

$$43 - [(4 \times 3) + (3 \times 4)] = 43 - 24 = 19$$

Step 6: Divide 19 by the flag pole: 19 ÷ 12 = 1, remainder 7. Put down 1 and put the remainder of 7 after the 3.

12^{34}	42	${}_{6}1_{4}3_{7}$
	3.	4 1

Solution: The answer is 3.41.

Altered Remainders

In some cases, you may get a negative when working a division problem using the flag method. What do you do? Let me introduce you something called *altered remainders*. With altered remainders, you decrease the quotient to increase the remainder.

For example, for 43 ÷ 8, the quotient is 5 and the remainder is 3. If you want to increase the remainder, you drop 1 from the quotient and add 8 to the remainder.

Quotient	Remainder
5	3
4	3 + 8 = 11
3	11 + 8 = 19
2	19 + 8 = 27
1	27 + 8 = 35

Let's take the same quotient and remainder and apply them to a different problem, 28 ÷ 5. In this case, you drop 1 from the quotient and add 5 to the remainder:

Quotient	Remainder
5	3
4	3 + 5 = 8
3	8 + 5 = 13
2	13 + 5 = 18
1	18 + 5 = 23

QUICK TIP

As you can see in the tables, how much you add to the remainder is not equal to how much you take away from the quotient. The change in the remainder depends on the value of the divisor.

The following examples show you how to apply altered remainders.

Example 1

Solve the problem 3,412 ÷ 24 to one decimal place.

Step 1: As you did for previous examples, lay out the problem. Because 3,412 is the dividend, put a line before it. For the divisor 24, put it to the left of the line, with the 2 serving as the flag pole and the 4 serving as the flag. You have one digit in the flag, so put a line one to the left of the dividend to indicate where the decimal point goes.

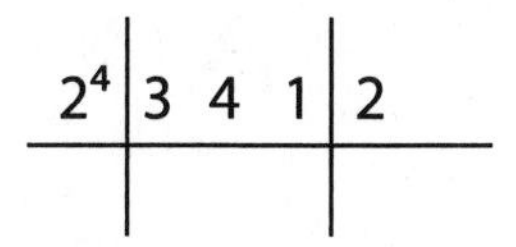

Step 2: Divide 3 by the flag pole, 2. This gives you a quotient of 1 and a remainder of 1 (because 2 × 1 = 2, which is the highest multiple of 2 that can go into 4). Put down 1 and prefix the remainder of 1 to 4 so it becomes 14.

2^4	$3\ {}_{1}4\ 1$	2
	1	

Step 3: Subtract 14 by the flag times the previous quotient digit.

14 – (4 × 1) = 14 – 4 = 10

Step 4: Divide 10 by the flag pole: 10 ÷ 2 = 5, remainder 0. Put down 5 and prefix the remainder of 0 to 1 so it becomes 01.

2^4	$3\ {}_{1}4\ {}_{0}1$	2
	1 5	

Step 5: Subtract 01 by the flag times the previous quotient digit: 01 – (4 × 5) = 1 – 20 = -20. Because you get a negative, reduce the previous quotient digit from 5 to 4; this increases the remainder from 0 to 2, which when prefixed to the 1 makes it 21.

2^4	$3\ {}_{1}4\ {}_{2}1$	2
	1 ~~5~~ 4	

Now subtract 21 by the flag times the previous quotient digit, which is now 4.

21 – (4 × 4) = 21 – 16 = 5

Step 6: Divide 5 by the flag pole: 5 ÷ 2 = 2, remainder 1. Put down 2 and prefix the remainder of 1 to 2 so it becomes 12.

2^4	$3\ {}_14\ {}_21$	${}_12$
	$1\ \not{5}\,4\,2$	

Step 7: Subtract 12 by the flag times the previous quotient digit.

$$12 - (4 \times 2) = 12 - 8 = 4$$

Step 8: Divide 4 by the flag pole: 4 ÷ 2 = 2, remainder 0. Put down 2 and prefix the remainder of 0 to 0 above so it becomes 00.

2^4	$3\ {}_14\ {}_21$	${}_12\ {}_00$
	$1\ \not{5}\,4\,2$	2

Step 9: Subtract 00 by the flag times the previous quotient digit: 00 – (4 × 2) = -8. Because you get a negative, reduce the previous quotient to 1, which changes the remainder from 0 to 2.

2^4	$3\ {}_14\ {}_21$	${}_12\ {}_20$
	$1\ \not{5}\,4\,2$	$\not{2}\ 1$

Solution: The answer is 142.1.

<u>**Example 2**</u>

Solve the problem 5,614 ÷ 21 to one decimal place.

Step 1: Lay out the problem. Because 5,614 is the dividend, put a line before it. For the divisor 21, put it to the left of the line, with the 2 serving as the flag pole and the 1 serving as the flag. You have one digit in the flag, so put a line one to the left of the dividend to indicate where the decimal point goes.

2^1	5 6 1	4

Step 2: Divide 5 by the flag pole, which is 2. This gives you a quotient of 2 with a remainder of 1 (because 2 × 2 = 4, which is the highest multiple of 2 that can go into 5). Put down 2 and prefix the remainder of 1 to 6 so it becomes 16.

2^1	$5\ {}_16\ 1$	4
	2	

Step 3: Subtract 16 by the flag times the previous quotient digit.

$16 - (1 \times 2) = 14$

Step 4: Divide 14 by the flag pole: 14 ÷ 2 = 7, remainder 0. Put down 7 as the quotient digit and prefix the remainder of 0 to 1 so it becomes 01.

2^1	$5\ {}_16\ {}_01$	4
	2 7	

Step 5: Subtract 01 by the flag times the previous quotient digit: 01 – (1 × 7) = -7. Because you get a negative, reduce the previous quotient from 7 to 6; this increases the remainder from 0 to 2, which prefixed to 1 becomes 21.

2^1	$5\ {}_16\ {}_21$	4
	2 ~~7~~ 6	

Now subtract 21 by the flag times the previous quotient digit, which is now 6.

$21 - (1 \times 6) = 15$

Step 6: Divide 15 by the flag pole: 15 ÷ 2 = 7, remainder 1. Put down 7 and prefix the remainder of 1 to 4 so it becomes 14.

2^1	$5\ {}_16\ {}_21$	${}_14$
	2~~7~~67.	

Step 7: Subtract 14 by the flag times the previous quotient digit

$14 - (1 \times 7) = 7$

Step 8: Divide 7 by the flag pole: 7 ÷ 2 = 3, remainder 1. Put down 3 and put the remainder of 1 after 4.

2^1	$5\ {}_16\ {}_21$	${}_14{}_10$
	2~~7~~67.	3

Solution: The answer is 267.3.

<u>**Example 3**</u>

Solve the problem 7,943 ÷ 42 to one decimal place.

Step 1: Lay out the problem. Because 7,943 is the dividend, put a line before it. For the divisor 42, put it to the left of the line, with the 4 serving as the flag pole and the 2 serving as the flag. You have one digit in the flag, so put a line one to the left of the dividend to indicate where the decimal point goes.

4^2	7 9 4	3

Step 2: Divide 7 by the flag pole, 4. This gives you a quotient digit of 1 and a remainder of 3 (because 4 × 1 = 4, which is the highest multiple of 4 that can go into 7). Put down 1 and prefix the remainder of 3 to 9 so it becomes 39.

4^2	$7\ {}_{3}9\ 4$	3
	1	

Step 3: Subtract 39 by the flag times the previous quotient digit.

$$39 - (2 \times 1) = 37$$

Step 4: Divide 37 by the flag pole: $37 \div 4 = 9$, remainder 1. Put down 9 and prefix the remainder of 1 to 4 so it becomes 14.

4^2	$7\ {}_{3}9\ {}_{1}4$	3
	$1\ 9$	

Step 5: Subtract 14 by the flag times the previous quotient digit: $14 - (2 \times 9) = 14 - 18 = -4$. Because you get a negative, bring down the previous quotient from 9 to 8; this increases the remainder from 1 to 5, which prefixed to 4 becomes 54.

4^2	$7\ {}_{3}9\ {}_{5}4$	3
	1 ~~9~~ 8	

Now subtract 54 by the flag times the previous quotient digit, which is now 8.

$$54 - (2 \times 8) = 54 - 16 = 38$$

Step 6: Divide 38 by the flag pole: $38 \div 4 = 9$, remainder 2. Put down 9 and prefix the remainder of 2 to 3 so it becomes 23.

4^2	$7\ {}_{3}9\ {}_{5}4$	${}_{2}3$
	1 ~~9~~ 8 9	

Step 7: Subtract 23 by the flag times the previous quotient digit.

$23 - (2 \times 9) = 23 - 18 = 5$

Step 8: Divide 5 by the flag pole: 5 ÷ 4 = 1, remainder 1. Put down 1 and put the remainder of 1 after 3.

4^2	$7\ {}_{3}9\ {}_{5}4$	${}_{2}3{}_{1}0$
	$1\ \not{9}\ 8\ 9.$	1

Solution: The answer is 189.1.

Auxiliary Fractions

Auxiliary fractions help you more easily find the decimal value of a fraction than doing long division. There are two types of auxiliary fractions: ones with a denominator ending in 9 or a series of 9s, and ones with a denominator ending in 1 or a series of 1s. The following sections go through each type. Once you have the process down, you'll have no trouble zipping through fractions!

Type 1: Fractions with a Denominator Ending in 9 or a Series of 9s

To find the auxiliary fraction for fractions in which the denominator ends in a 9 or a series of 9s, you drop the 9 or 9s from the denominator and increase the remaining number by 1; you then divide the numerator by 10. After that, you begin dividing.

For each remainder, you prefix it to the previous quotient and then divide that by the denominator until you get the answer to the number of decimal places you require.

<u>Example 1</u>

Solve the problem $\frac{6}{49}$ to four decimal places.

Step 1: Find the auxiliary fraction. For $\frac{6}{49}$, drop the 9 from the denominator and increase 4 by 1 so the denominator is now 5. Finish by dividing the numerator by 10: 6 ÷ 10 = 0.6. This gives you an auxiliary fraction of $\frac{0.6}{5}$.

Step 2: Divide 0.6 by 5. This gives you a quotient of 0.1 and a remainder of 1. Put down the 0.1 and put the remainder of 1 in front of it so it becomes 1.1. This is the next dividend.

$$\frac{6}{49} = \frac{0.6}{5} \quad \text{AF} = 0.{}_{1}1$$

Step 3: Ignoring the decimal, divide 1.1 by 5: 11 ÷ 5 = 2, remainder 1. Put down 2 and put the remainder of 1 in front of it so it becomes 12. This is the next dividend.

$$\frac{6}{49} = \frac{0.6}{5} \quad \text{AF} = 0.{}_{1}1{}_{1}2$$

Step 4: Divide 12 by 5: 12 ÷ 5 = 2, remainder 2. Put down 2 and put the remainder of 2 in front of it so it becomes 22. This is the next dividend.

$$\frac{6}{49} = \frac{0.6}{5} \quad \text{AF} = 0.{}_{1}1{}_{1}2{}_{2}2$$

Step 5: Divide 22 by 5: 22 ÷ 5 = 4, remainder 2. Put down the 4 and put the remainder of 2 in front of it. Because you're only finding the answer to four decimal places, you don't need to continue.

$$\frac{6}{49} = \frac{0.6}{5} \quad \text{AF} = 0.{}_{1}1{}_{1}2{}_{2}2{}_{2}4\ldots$$

Solution: The answer is 0.1224.

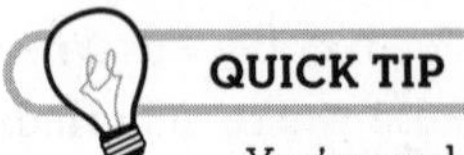

You're probably getting tired of reading "This is the next dividend" at the end of each step. However, I think it's a good way to emphasize that every number you get is used to get the next one. If you have other tricks to keep that idea in your head, feel free to use them. It's all about making this process easier for you!

Example 2

Solve the problem $\frac{11}{149}$ to four decimal places.

Step 1: Find the auxiliary fraction. For $\frac{11}{149}$, drop the 9 from the denominator and increase 14 by 1 so the denominator is now 15. Finish by dividing the numerator by 10: 11 ÷ 10 = 1.1. This gives you an auxiliary fraction of $\frac{1.1}{15}$.

Step 2: Ignoring the decimal, divide 1.1 by 15. This gives you a quotient of 0 and a remainder of 11. Put down 0 and put the remainder of 11 before it so it becomes 110. This is the next dividend.

$$\frac{11}{149} = \frac{1.1}{15} \quad AF = 0._{11}0$$

Step 3: Divide 110 by 15: 110 ÷ 15 = 7, remainder 5. Put down 7 and put the remainder of 5 before it so it becomes 57. This is the next dividend.

$$\frac{11}{149} = \frac{1.1}{15} \quad AF = 0._{11}0_{5}7$$

Step 4: Divide 57 by 15: 57 ÷ 15 = 3, remainder 12.Put down 3 and put the remainder of 12 before it so it becomes 123. This is the next dividend.

$$\frac{11}{149} = \frac{1.1}{15} \quad AF = 0._{11}0_{5}7_{12}3$$

Step 5: Divide 123 by 15: 123 ÷ 15 = 8, remainder 3. Put down 8 and put the remainder of 3 before it. Because you're only finding the answer to four decimal places, you don't need to continue.

$$\frac{11}{149} = \frac{1.1}{15} \text{ AF} = 0._{11}0_{5}7_{12}3_{3}8...$$

Solution: The answer is 0.0738.

<u>**Example 3**</u>

Solve the problem $\frac{16}{19}$ to four decimal places.

Step 1: Find the auxiliary fraction. For $\frac{16}{19}$, drop the 9 from the denominator and increase 1 by 1 so the denominator is now 2. Finish by dividing the numerator by 10: 16 ÷ 10 = 1.6. This gives you an auxiliary fraction of $\frac{1.6}{2}$.

Step 2: Ignoring the decimal, divide 1.6 by 2. This gives you a quotient of 8 and a remainder of 0. Put down 8 and put the remainder of 0 before it so it becomes 08. This is the next dividend.

$$\frac{16}{19} = \frac{1.6}{2} \text{ AF} = 0._{0}8$$

Step 3: Divide 08 by 2: 08 ÷ 2 = 4, remainder 0. Put down 4 and put the remainder of 0 before it so it becomes 04. This is the next dividend.

$$\frac{16}{19} = \frac{1.6}{2} \text{ AF} = 0._{0}8_{0}4$$

Step 4: Divide 04 by 2: 04 ÷ 2 = 2, remainder 0. Put down 2 and put the remainder of 0 before it so it becomes 02. This is the next dividend.

$$\frac{16}{19} = \frac{1.6}{2} \text{ AF} = 0._{0}8_{0}4_{0}2$$

Step 5: Divide 02 by 2: 02 ÷ 2 = 1, remainder 0. Put down 1 and put the remainder of 0 before it. Because you're only finding the answer to four decimal places, you don't need to continue.

$$\frac{16}{19} = \frac{1.6}{2}\ \text{AF} = 0._{0}8_{0}4_{0}2_{0}1\ldots$$

Solution: The answer is 0.8421.

Type 2: Fractions with a Denominator Ending in 1 or a Series of 1s

To find the auxiliary fraction for fractions in which the denominator ends in a 1 or a series of 1s, the numerator and denominator are both reduced by 1 and the top and bottom are divided by 10.

You get the next dividend by first writing down the remainder and then finding the difference between 9 and the quotient. The first part of the dividend is the remainder, while the second part is the difference.

Example 1

Solve the problem $\frac{4}{21}$ to four decimal places.

Step 1: Find the auxiliary fraction. For $\frac{4}{21}$, reduce the numerator and denominator by 1, making it $\frac{3}{20}$. Finish by dividing the numerator and denominator by 10: 3 ÷ 10 = 0.3; 20 ÷ 10 = 2. This gives you an auxiliary fraction of $\frac{0.3}{2}$.

Step 2: Ignoring the decimal, divide 0.3 by 2. This gives you a quotient of 1 and a remainder of 1. Put down 1 and put the remainder of 1 before it.

$$0._{1}1$$

Step 3: Subtract the quotient from 9: 9 – 1 = 8. Put the 8 on top of the 1. The remainder and this answer are the next dividend: 18.

$$\frac{4}{21} = \frac{3}{20}\ \text{AF} = \frac{0.3}{2}\ \text{AF} = 0.{}_1\overset{8}{1}$$

Step 4: Divide 18 by 2: 18 ÷ 2 = 9, remainder 0. Put down 9 and put the remainder of 0 before it.

$$\frac{4}{21} = \frac{3}{20}\ \text{AF} = \frac{0.3}{2}\ \text{AF} = 0.{}_1\overset{8}{1}{}_0 9$$

Step 5: Subtract the quotient from 9: 9 – 9 = 0. Put the 0 on top of the 9. The remainder and this answer are the next dividend: 00.

$$\frac{4}{21} = \frac{3}{20}\ \text{AF} = \frac{0.3}{2}\ \text{AF} = 0.{}_1\overset{8}{1}{}_0\overset{0}{9}$$

Step 6: Divide 00 by 2: 00 ÷ 2 = 0, remainder 0. Put down 0 and put the remainder of 0 before it.

$$\frac{4}{21} = \frac{3}{20}\ \text{AF} = \frac{0.3}{2}\ \text{AF} = 0.{}_1\overset{8}{1}{}_0\overset{0}{9}{}_2 0$$

Step 7: Subtract the quotient from 9: 9 – 0 = 9. Put the 9 on top of 0. The remainder and this answer are the next dividend: 09.

$$\frac{4}{21} = \frac{3}{20}\ \text{AF} = \frac{0.3}{2}\ \text{AF} = 0.{}_1\overset{8}{1}{}_0\overset{0}{9}{}_2\overset{9}{0}$$

Step 8: Divide 09 by 2: 09 ÷ 2 = 4, remainder 1. Put down 4 and put the remainder of 1 before it.

$$\frac{4}{21} = \frac{3}{20}\ \text{AF} = \frac{0.3}{2}\ \text{AF} = 0.{}_1\overset{8}{1}{}_0\overset{0}{9}{}_2\overset{9}{0}{}_1 4$$

Step 9: Subtract the quotient from 9: 9 – 4 = 5. Put the 5 on top of the 4.

$$\frac{4}{21} = \frac{3}{20}\ \text{AF} = \frac{0.3}{2}\ \text{AF} = 0.{}_1\overset{8}{1}{}_0\overset{0}{9}{}_2\overset{9}{0}{}_1\overset{5}{4}$$

Solution: The answer is 0.1904.

<u>Example 2</u>

Solve the problem $\frac{8}{31}$ to four decimal places.

Step 1: Find the auxiliary fraction. For $\frac{8}{31}$, reduce the numerator and denominator by 1, making it $\frac{7}{30}$. Finish by dividing the numerator and denominator by 10: 7 ÷ 10 = 0.7; 30 ÷ 10 = 3. This gives you an auxiliary fraction of $\frac{0.7}{3}$.

Step 2: Ignoring the decimal, divide 0.7 by 3. This gives you a quotient of 2 and a remainder of 1. Put down 2 and put the remainder of 1 before it.

$$\frac{8}{31} = \frac{7}{30} \quad AF = \frac{0.7}{3} \quad AF = 0.{}_{1}2$$

Step 3: Subtract the quotient from 9: 9 – 2 = 7. Put the 7 on top of the 2. The remainder and this answer are the next dividend: 17.

$$\frac{8}{31} = \frac{7}{30} \quad AF = \frac{0.7}{3} \quad AF = 0.{}_{1}2^{7}$$

Step 4: Divide 17 by 3: 17 ÷ 3 = 5, remainder 2. Put down 5 and put the remainder of 2 before it.

$$\frac{8}{31} = \frac{7}{30} \quad AF = \frac{0.7}{3} \quad AF = 0.{}_{1}2^{7}{}_{2}5$$

Step 5: Subtract the quotient from 9: 9 – 5 = 4. Put the 4 on top of the 5. The remainder and this answer are the next dividend: 24.

$$\frac{8}{31} = \frac{7}{30} \quad AF = \frac{0.7}{3} \quad AF = 0.{}_{1}2^{7}{}_{2}5^{4}$$

Step 6: Divide 24 by 3: 24 ÷ 3 = 8, remainder 0. Put down 8 and put the remainder of 0 before it.

$$\frac{8}{31} = \frac{7}{30} \quad AF = \frac{0.7}{3} \quad AF = 0.{}_{1}2^{7}{}_{2}5^{4}{}_{0}8$$

Step 7: Subtract the quotient from 9: 9 – 8 = 1. Put the 1 on top of the 8. The remainder and this answer are the next dividend: 01.

$$\frac{8}{31} = \frac{7}{30}\ AF = \frac{0.7}{3}\ AF = 0.{}_{1}2{}^{7}{}_{2}5{}^{4}{}_{0}8^{1}$$

Step 8: Divide 01 by 3: 01 ÷ 3 = 0, remainder 1. Put down 0 and put the remainder of 1 before it.

$$\frac{8}{31} = \frac{7}{30}\ AF = \frac{0.7}{3}\ AF = 0.{}_{1}2{}^{7}{}_{2}5{}^{4}{}_{0}8^{1}{}_{1}0$$

Step 9: Subtract the quotient from 9: 9 – 0 = 9. Put the 9 on top of the 0.

$$\frac{8}{31} = \frac{7}{30}\ AF = \frac{0.7}{3}\ AF = 0.{}_{1}2{}^{7}{}_{2}5{}^{4}{}_{0}8^{1}{}_{1}0^{9}$$

Solution: The answer is 0.2580.

The Least You Need to Know

- To do the flag method, split the divisor into a "flag pole" and "flag," and then follow two simple rules: divide and subtract.
- Altered remainders help you avoid a negative number. You simply decrease the quotient and increase the remainder.
- If you want to find the decimal value of a fraction, you can use an auxiliary fraction. You alter the fraction based on whether it ends in 1 or 9 and solve.

CHAPTER

5

Checking Your Answers with Digit Sums

In This Chapter

- Digit sums explained
- Casting out nines to further simplify your calculations
- Creating a nine-point circle and understanding how it works
- Checking your addition, subtraction, multiplication, and division answers using digit sums

In previous chapters, you learned speedier ways to solve addition, subtraction, multiplication, and division problems. But how do you know if they're right?

In this chapter, I talk about the concept of digit sums, which can help you check your calculations in a simple—not to mention quick—manner.

What Are Digit Sums?

The word *digit* means numbers like 1, 2, 3, 4, and so on, while and the word *sum* means to "add." So basically, to find the digit sum of a number, you simply have to add the digits in the numbers until you get a single-digit number.

For example, if you have to find the digit sum of 61, you simply add each digit of the number together: $6 + 1 = 7$. So the digit sum of 61 is 7.

If you end up with a double-digit number, you continue to add the individual digits until you get a single-digit number. For example, when you add the digits for the number 538, you end up with 16 ($5 + 3 + 8$). To get a single-digit number, add again: $1 + 6 = 7$. For 538, the digit sum is 7.

The following are some other examples of digit sums.

Number	Adding the Digits	Digit Sum
65	$6 + 5 = 11$; $1 + 1 = 2$	2
721	$7 + 2 + 1 = 10$; $1 + 0 = 1$	1
3,210	$3 + 2 + 1 + 0 = 6$	6
67,754	$6 + 7 + 7 + 5 + 4 = 29$; $2 + 9 = 11$; $1 + 1 = 2$	2

As you can see, this method is simple and very easy to understand. However, you can use some shortcuts to make the process even quicker.

Casting Out Nines

To make the digit-sum process even simpler, you can use a method called *casting out nines*. For this method, when trying to find the digit sum of a number, you first cast out or ignore any 9s and numbers adding up to 9. After that, you add up and reduce the remaining digits to a single-digit number to get the digit sum.

Let's look at an example. Say you have to find the digit sum of 8,154,912,320. Instead of adding all those digits together, you can cast out the 9s and any numbers that add to 9. For this number, you cast out the 9, the 8 and 1, and the 5 and 4, because they are or add up to 9. After they're taken out, the number looks like this:

~~81549~~12320

You then add the remaining digits to get the digit sum:

$$1 + 2 + 3 + 2 + 0 = 8$$

The digit sum of 8,154,912,320 is 8.

Let's try this method again using the number 970,230,612. In this number, you can cast out the 9, the 7 and 2, and the 3 and 6, as all of these are or add to 9. Afterward, the number looks like the following:

~~97~~0~~23~~0~~6~~12

You now add the remaining digits:

$$0 + 0 + 1 + 2 = 3$$

The digit sum for 970,230,612 is 3.

The Nine-Point Circle

The nine-point circle is a valuable aid when reducing numbers to their digit sums. To create a nine-point circle, you need to first list the numbers, followed by their digit sum. The digit cycles 1 through 9, as anything higher than 9 would have to be reduced. In the following table, I've listed 1 to 18 and the digit sums associated with them.

Number	1	2	3	4	5	6	7	8	9	10	11	12	13	14	15	16	17	18
Digit Sums	1	2	3	4	5	6	7	8	9	1	2	3	4	5	6	7	8	9

Here's how these values look when placed on a nine-point circle.

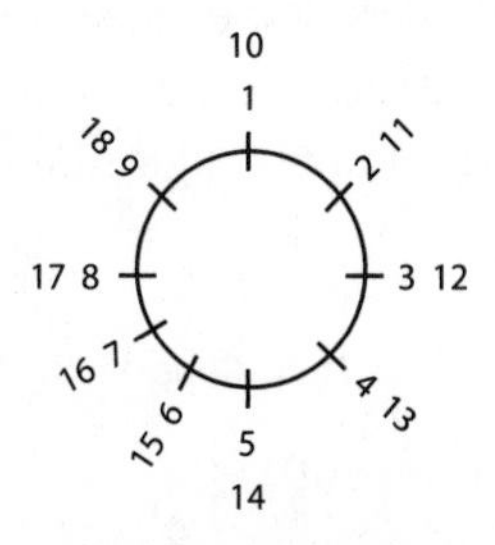

The nine-point circle.

The point to note here is that numbers on the same branch of a nine-point circle have the same digit sum—1 and 10 have a digit sum of 1, 2 and 11 have a digit sum of 2, and so on. Also, if you add or take away 9 from any of the numbers, the digit sum remains the same.

You can build out the branches as far as you need to in order to help you remember what certain numbers reduce to. For example, for a 1 branch, you could expand out beyond 1 and 10 to include 19, 28, 37, and on up.

QUICK TIP

In the nine-point circle, the digit sums 9 and 0 are considered to be equivalent. This is because if you add 9 to 9, you get 18, which has a digit sum of 9; if you take away 9 from 9, it gives you 0. So you can say that in 8,136, the digit sum is 9 or 0. Both answers would be correct!

Using Digit Sums to Check Your Answers

You know what digit sums are and the different ways you can find them. Now, with the help of this concept, you can check your addition, subtraction, multiplication, and division answers. I show the long form for calculating the digit sums in the following sections, but feel free to use casting out nines and the nine-point circle to shorten the work.

Checking Addition Answers

To check your addition using digit sums, you simply reduce each number and then make sure the digit sums of the problem numbers added together are equal to the digit sum of the answer.

<u>Example 1</u>

Verify that the answer for the addition problem 734 + 352 is 1,086.

Step 1: Find the digit sum of 734.

$7 + 3 + 4 = 14$
$1 + 4 = 5$

Step 2: Find the digit sum of 352.

$3 + 5 + 2 = 10$
$1 + 0 = 1$

Step 3: Add the two digit sums together.

$5 + 1 = 6$

Step 4: Find the digit sum of 1,086.

$1 + 0 + 8 + 6 = 15$
$1 + 5 = 6$

Step 5: Check the digit sums against each other.

$$\begin{array}{r} 734 \\ +352 \\ \hline 1086 \end{array} \quad \begin{array}{c} \rightarrow \\ \rightarrow \\ \rightarrow \end{array} \quad \begin{array}{r} 5 \\ +1 \\ \hline 6 \end{array}$$

Solution: Both are 6, meaning the answer is correct.

Example 2

Verify the answer for the addition problem 2,344 + 6,235 is 8,579.

Step 1: Find the digit sum of 2,344.

$2 + 3 + 4 + 4 = 13$
$1 + 3 = 4$

Step 2: Find the digit sum of 6,235.

$6 + 2 + 3 + 5 = 16$
$1 + 6 = 7$

Step 3: Add the two digit sums.

$4 + 7 = 11$
$1 + 1 = 2$

Step 4: Find the digit sum of 8,579.

$8 + 5 + 7 + 9 = 29$
$2 + 9 = 11$
$1 + 1 = 2$

Step 5: Make sure the combined digit sums of the problem numbers match the digit sum of the answer.

$$\begin{array}{r} 2344 \\ +\,6235 \\ \hline 8579 \end{array} \begin{array}{c} \rightarrow \\ \rightarrow \\ \rightarrow \end{array} \begin{array}{r} 4 \\ +\,7 \\ \hline 11 = 1+1 = 2 \end{array}$$

Solution: Both are 2, meaning the answer is absolutely correct!

Checking Subtraction Answers

Checking your subtraction answers using digit sums is very similar to how you checked your addition answers, except you subtract instead of add.

QUICK TIP

Remember, if you don't get a single-digit number after you add the digits, continue adding until you do. For example, if you get 15 in the first round of addition, add 1 and 5.

<u>Example 1</u>

Verify that the answer to the subtraction problem 74,637 – 24,267 is 50,370.

Step 1: Find the digit sum of 74,637.

$7 + 4 + 6 + 3 + 7 = 27$
$2 + 7 = 9$

Step 2: Find the digit sum of 24,267.

$2 + 4 + 2 + 6 + 7 = 21$
$2 + 1 = 3$

Step 3: Subtract the two digit sums.

$9 - 3 = 6$

Step 4: Find the digit sum of 50,370.

$5 + 3 + 7 = 15$
$1 + 5 = 6$

Step 5: Make sure the two digit sums are the same.

$$\begin{array}{r} 74637 \\ -\,24267 \\ \hline 50370 \end{array} \begin{array}{c} \rightarrow \\ \rightarrow \\ \rightarrow \end{array} \begin{array}{r} 9 \\ -\,3 \\ \hline 6 \end{array}$$

Solution: Both are 6, so you can safely say the answer is correct.

<u>Example 2</u>

Verify that the answer for the subtraction problem 4,321 – 1,786 is 2,535.

Step 1: Find the digit sum of 4,321.

$4 + 3 + 2 + 1 = 10$
$1 + 0 = 1$

Step 2: Find the digit sum of 1,786.

$1 + 7 + 8 + 6 = 22$
$2 + 2 = 4$

Step 3: Subtract the two digit sums.

$1 - 4 = -3$

Step 4: The digit sum for the two values can't be negative, because the answer's digit sum won't be a negative number; to make it positive, subtract –3 from 9.

$9 - 3 = 6$

Step 5: Find the digit sum of 2,535.

$2 + 5 + 3 + 5 = 15$
$1 + 5 = 6$

Step 6: Check the two digit sums against each other to see if they match.

$$\begin{array}{r} 4321 \\ -\,1786 \\ \hline 2535 \end{array} \begin{array}{c} \rightarrow \\ \rightarrow \\ \rightarrow \end{array} \begin{array}{r} 1 \\ -\,4 \\ \hline -3 = 9 - 3 = 6 \end{array}$$

Solution: Both are 6, which means the answer is correct.

Checking Multiplication Answers

Using digit sums to check multiplication answers involves multiplying the digit sums of the problem numbers to see if they match the answer. If you need to reduce the number, though, you add the digits together.

Example 1

Verify that the answer for the multiplication problem 62×83 is 5,146.

Step 1: Find the digit sum of 62.

$$6 + 2 = 8$$

Step 2: Find the digit sum of 83.

$$8 + 3 = 11$$
$$1 + 1 = 2$$

Step 3: Multiply the two digit sums. In this case, you have to reduce, which means you add the digits of the answer to get the single-digit number.

$$8 \times 2 = 16$$
$$1 + 6 = 7$$

Step 4: Find the digit sum of 5,146.

$$5 + 1 + 4 + 6 = 16$$
$$1 + 6 = 7$$

Step 5: Check the digit sums against each other to see if they match.

$$\begin{array}{rcr} 62 & \rightarrow & 8 \\ \times 83 & \rightarrow & \times 2 \\ \hline 5146 & \rightarrow & 16 = 1 + 6 = 7 \end{array}$$

Solution: Both are 7; therefore, you can conclude your answer is correct.

Example 2

Verify that the answer for the multiplication problem 726 × 471 is 341,946.

Step 1: Find the digit sum of 726.

$7 + 2 + 6 = 15$
$1 + 5 = 6$

Step 2: Find the digit sum of 471.

$4 + 7 + 1 = 12$
$1 + 2 = 3$

Step 3: Multiply the two digit sums and add the answer to reduce it to a single digit.

$6 \times 3 = 18$
$1 + 8 = 9$

Step 4: Find the digit sum of 341,946.

$3 + 4 + 1 + 9 + 4 + 6 = 27$
$2 + 7 = 9$

Step 5: Check the digit sums to see if they're the same.

$$\begin{array}{rcr} 726 & \rightarrow & 6 \\ \times 471 & \rightarrow & \times 3 \\ \hline 341946 & \rightarrow & 18 = 1 + 8 = 9 \end{array}$$

Solution: Both are 9, so the answer is absolutely correct!

SPEED BUMP

The digit-sum method is a great tool for checking your answers, but it has its limitations. For example, if you think the sum of 12 × 34 is 804 instead of 408, you'll get the same digit sum (3), but the answer will be wrong.

$$\begin{array}{rcr} 12 & \rightarrow & 3 \\ \times 34 & \rightarrow & \times 7 \\ \hline 804 & \rightarrow & 21 = 2 + 1 = 3 \end{array}$$

Wrong Answer

So keep this in mind and be a little careful when using this checking tool.

Checking Division Answers

When doing a digit sums check for a division problem, the digit sum of the divisor times the digit sum of the answer should be equal to the digit sum of the dividend.

Example 1

Verify that the answer for 112 ÷ 7 is 16.

Step 1: Find the digit sum of the divisor, 7. In this case, it's simply 7.

Step 2: Find the digit sum of the answer.

$1 + 6 = 7$

Step 3: Multiply the digit sum of the divisor and the answer, and reduce to a single digit.

$7 \times 7 = 49$
$4 + 9 = 13$
$1 + 3 = 4$

Step 4: Find the digit sum of the dividend, 112.

$1 + 1 + 2 = 4$

Step 5: Compare the digit sum of the combined answer and divisor with the digit sum of the dividend to see if they're the same.

Digit sum of divisor and answer: 4
Digit sum of dividend: 4

Solution: Both are 4, so the answer is correct.

Example 2

Verify that 464 ÷ 8 is 58.

Step 1: Find the digit sum of the divisor, 8. In this case, it's simply 8.

Step 2: Find the digit sum of the answer.

$5 + 8 = 13$
$1 + 3 = 4$

Step 3: Multiply the digit sum of the divisor and the answer, and reduce to a single digit.

$8 \times 4 = 32$
$3 + 2 = 5$

Step 4: Find the digit sum of the dividend, 464.

$4 + 6 + 4 = 14$
$1 + 4 = 5$

Step 5: Compare the digit sum of the combined answer and divisor with the digit sum of the dividend to see if they're the same.

Digit sum of divisor and answer: 5
Digit sum of dividend: 5

Solution: Because the two digit sums match, the answer is correct.

The Least You Need to Know

- To find the digit sum, add and reduce the digits of a number until you get a single digit.
- Casting out nines and using a nine-point circle as guidance can help you eliminate unnecessary steps when calculating digit sums.
- For addition, subtraction, and multiplication problems, an answer is correct if the digit sum of the problem numbers combined is equal to the digit sum of the answer.
- For division problems, the answer is correct if the digit sum of the combined divisor and answer is equal to the digit sum of the dividend.

PART

2

High-Speed Math Applications

When you encounter math with decimals, fractions, or square roots, you're probably tempted to pull out a calculator and let it do the work for you. But what if you don't have a calculator on hand? Not to worry—speed math has ways to deal with those and more. Over time, you may even forgo a pencil and paper to get your answers.

In this part, I teach you speed math tips and methods for fractions, decimals, percentages, divisibility, squared numbers, cubed numbers, numbers to the fourth or fifth power, and square and cube roots.

CHAPTER

6

Fractions

In This Chapter

- Learning how to add and subtract fractions
- Learning how to multiply and divide fractions
- Applying the vertically and crosswise method to fractions

You may be one of the many people who doesn't enjoy math using fractions. However, once you have a grasp of the many quick and easy ways you can work fraction problems, you'll gain a comfort level even with the most complex-looking fraction problems.

In this chapter, you learn ways to add, subtract, multiply, and divide fractions, as well as how you can use the vertically and crosswise method when working fractions.

Adding Fractions

Adding fractions can be tricky to master, particularly if the denominators don't match. In the following sections, I provide the fastest, simplest ways to solve these problems.

Fractions with the Same Denominator

Whenever two or more fractions have the same denominator, you add the numerators and simplify the answer as much as possible.

Example 1

Solve the problem $\frac{5}{11} + \frac{3}{11}$.

Because the denominators are same, you simply add the numerators.

$$\frac{5+3}{11} = \frac{8}{11}$$

Solution: The answer is $\frac{8}{11}$.

Example 2

Solve the problem $\frac{3}{10} + \frac{1}{10}$.

Here, too, the denominators are same, so you just add the numerators and simplify the fraction to get your answer.

$$\frac{3+1}{10} = \frac{4}{10} = \frac{2}{5}$$

Solution: The answer is $\frac{2}{5}$.

Fractions Where One Denominator Is a Factor of the Other

To add fractions that don't have matching denominators, but where one is a factor of the other, you multiply the numerator and denominator by that factor and solve.

Example 1

Solve the problem $\frac{2}{5} + \frac{9}{20}$.

Step 1: Because 5 is a factor of 20, multiply the numerator and denominator of $\frac{2}{5}$ by 4 to get the fraction you can use.

$$\frac{2}{5} \times \frac{4}{4} = \frac{8}{20}$$

Step 2: Add the numerators of the fractions.

$$\frac{8}{20} + \frac{9}{20} = \frac{17}{20}$$

Solution: The answer is $\frac{17}{20}$.

<u>Example 2</u>

Solve the problem $\frac{11}{15} + \frac{7}{30}$.

Step 1: Because 15 is a factor of 30, multiply the numerator and denominator of $\frac{11}{15}$ by 2 to get the fraction you can use.

$$\frac{11}{15} \times \frac{2}{2} = \frac{22}{30}$$

Step 2: Add the numerators of the fractions.

$$\frac{22+7}{30} = \frac{29}{30}$$

Solution: The answer is $\frac{29}{30}$.

Addition with the Vertically and Crosswise Method

To add fractions without a common denominator or one that's a factor of the other, you can use the vertically and crosswise method.

<u>Example 1</u>

Solve the problem $\frac{3}{5} + \frac{1}{4}$.

Step 1: Multiply the fractions crosswise and add the products: $(4 \times 3) + (5 \times 1) = 17$. This is the numerator.

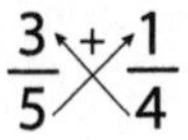

$$\frac{3}{5} \times \frac{1}{4}$$

Step 2: Multiply the denominators: 5 × 4 = 20. This is the denominator.

$$\frac{3}{5} + \frac{1}{4}$$

Solution: The answer is $\frac{17}{20}$.

<u>Example 2</u>

Solve the problem $\frac{2}{11} + \frac{7}{9}$.

Step 1: Multiply the fractions crosswise and add the products: (9 × 2) + (11 × 7) = 18 + 77 = 95. This is the numerator.

$$\frac{2}{11} + \frac{7}{9}$$

Step 2: Multiply the denominators: 11 × 9 = 99. This is the denominator.

$$\frac{2}{11} + \frac{7}{9}$$

Solution: The answer is $\frac{95}{99}$.

Subtracting Fractions

Subtracting fractions is very similar to what you learned for adding fractions—the main difference is that a minus sign is involved. The following show you different ways you can subtract, based on the setup of the denominators.

Fractions with the Same Denominator

This is the easiest type of subtraction for fractions. Whenever you spot that the denominators are same, you can just subtract the numerators and write your answer.

Example 1

Solve the problem $\frac{15}{16} - \frac{3}{16}$.

Because the denominators are the same, you simply subtract the numerators and simplify the fraction.

$$\frac{15-3}{16} = \frac{12}{16} = \frac{3}{4}$$

Solution: The answer is $\frac{3}{4}$.

SPEED BUMP

If you can simplify the fraction, do so; the fraction should be in the simplest form possible. Otherwise, you'll have an incomplete answer.

Example 2

Solve the problem $\frac{13}{35} - \frac{6}{35}$.

Because the denominators are the same, subtract the numerators and simplify.

$$\frac{13}{35} - \frac{6}{35} = \frac{7}{35} = \frac{1}{5}$$

Solution: The answer is $\frac{1}{5}$.

Fractions Where One Denominator Is a Factor of the Other

As you learned in the addition section, fractions where one denominator is a factor of the other require you to multiply the numerator and denominator by that factor before you can solve the problem.

Example 1

Solve the problem $\frac{1}{2} - \frac{5}{24}$.

Step 1: Because 2 is a factor of 24, multiply the numerator and denominator of $\frac{1}{2}$ by 12 to get the fraction you can use.

$$\frac{1}{2} \times \frac{12}{12} = \frac{12}{24}$$

Step 2: Subtract the numerators of the fractions.

$$\frac{12}{24} - \frac{5}{24} = \frac{7}{24}$$

Solution: The answer is $\frac{7}{24}$.

Example 2

Solve the problem $\frac{7}{9} - \frac{2}{3}$.

Step 1: Because 3 is a factor of 9, multiply the numerator and denominator of $\frac{2}{3}$ by 3 to get the fraction you can use.

$$\frac{2}{3} \times \frac{3}{3} = \frac{6}{9}$$

Step 2: Subtract the numerators of the fractions.

$$\frac{7}{9} - \frac{6}{9} = \frac{1}{9}$$

Solution: The answer is $\frac{1}{9}$.

Subtraction with the Vertically and Crosswise Method

Just like in addition, the vertically and crosswise method plays an important role in subtraction. You can subtract any fraction from any other fraction, no matter the denominator, using this method.

Example 1

Solve the problem $\frac{6}{7} - \frac{1}{2}$.

Step 1: Multiply the fractions crosswise and add the products: $(6 \times 2) - (7 \times 1) = 12 - 7 = 5$. This is the numerator.

$$\frac{6}{7} \times \frac{1}{2}$$

Step 2: Multiply the denominators: $7 \times 2 = 14$. This is the denominator.

$$\frac{6}{7} - \frac{1}{2}$$

Step 3: Subtract the fractions.

$$\frac{12 - 7}{14} = \frac{5}{14}$$

Solution: The answer is $\frac{5}{14}$.

Example 2

Solve the problem $\frac{12}{25} - \frac{3}{50}$.

Step 1: Multiply the fractions crosswise and add the products: $(12 \times 50) - (25 \times 3) = 600 - 75 = 525$. This is the numerator.

$$\frac{12}{25} \times \frac{3}{50}$$

Step 2: Multiply the denominators: 25 × 50 = 1,250. This is the denominator.

$$\frac{12}{25} - \frac{3}{50}$$

Step 3: Subtract the fractions.

$$\frac{600}{1250} - \frac{75}{1250} = \frac{525}{1250}$$

Step 4: The numerator and denominator are multiples of 25, so reduce the fraction by 25.

$$\frac{525}{1250} \div \frac{25}{25} = \frac{21}{50}$$

Solution: The answer is $\frac{21}{50}$.

Multiplying Fractions

If you need to multiply two fractions, all you need to do is multiply the numerators and then multiply the denominators.

<u>Example 1</u>

Solve the problem $\frac{4}{11} \times \frac{6}{19}$.

Step 1: Multiply the numerators: 4 × 6 = 24.

$$\frac{4}{11} \times \frac{6}{19} = \frac{24}{}$$

Step 2: Multiply the denominators: 11 × 19 = 209.

$$\frac{4}{11} \times \frac{6}{19} = \frac{24}{209}$$

Solution: The answer is $\frac{24}{209}$.

Example 2

Solve the problem $3\frac{4}{7} \times 6\frac{2}{3}$.

Step 1: Make the whole numbers part of the fractions. For $3\frac{4}{7}$, integrate the 3 by multiplying it by the denominator and adding it to the numerator: 3 × 7 = 21 + 4 = 25. This makes the first fraction $\frac{25}{7}$. For $6\frac{2}{3}$, integrate the 6 by multiplying it by denominator and adding it to the numerator: 6 × 3 = 18 + 2 = 20. This makes the second fraction $\frac{20}{3}$.

Step 2: Multiply the numerators and denominators. For the numerators: 25 × 20 = 500. For the denominators: 7 × 3 = 21.

$$\frac{25}{7} \times \frac{20}{3} = \frac{500}{21}$$

Step 3: Convert it to the mixed number form again: 500 ÷ 21 = 23, remainder 17. The 17 is the numerator.

$$23\frac{17}{21}$$

Solution: The answer is $23\frac{17}{21}$.

Dividing Fractions

Dividing fractions can be a headache. However, it doesn't have to be challenging—you can get the correct answer by multiplying the first fraction by the reciprocal of the second fraction.

Example 1

Solve the problem $\frac{9}{16} \div \frac{72}{32}$.

Step 1: Change the division sign to a multiplication sign and reverse the second fraction so 72 is on top and 32 is on the bottom.

$$\frac{9}{16} \times \frac{32}{72}$$

Step 2: Simplify the fractions crosswise and multiply. In this case, 32 goes into 16 twice, so 16 becomes 1 and 32 becomes 2; 72 goes into 9 eight times, so 9 becomes 1 and 72 becomes 8. You can then further reduce when you get the answer.

$$\frac{9}{16} \times \frac{32}{72} = \frac{1}{4}$$

Solution: The answer is $\frac{1}{4}$.

QUICK TIP

Whether it's on paper or in your head, right after you change the sign from division to multiplication, flip the numerator and denominator on the second fraction. That way, you'll keep those two processes connected.

Example 2

Solve the problem $\frac{18}{45} \div \frac{7}{90}$.

Step 1: Change the division sign to a multiplication sign and reverse the second fraction so 7 is on top and 90 is on the bottom.

$$\frac{18}{45} \times \frac{90}{7}$$

Step 2: Simplify the fractions crosswise and multiply. In this case, 90 goes into 45 twice, so 45 becomes 1 and 90 becomes 2; 18 and 7 can't reduce.

$$\frac{18}{45} \times \frac{90}{7} = \frac{18}{1} \times \frac{2}{7} = \frac{36}{7}$$

Step 3: Because the numerator is larger than the denominator, convert it to a mixed number: 36 ÷ 7 = 5, remainder 1. The 1 is the numerator.

$$\frac{36}{7} = 5\frac{1}{7}$$

Solution: The answer is $5\frac{1}{7}$.

The Least You Need to Know

- If the fractions in the problem have the same denominator, you can simply solve and reduce if necessary—no extra steps are involved.
- For fractions in which one denominator is a factor of the other, you multiply the numerator and denominator by that factor to get the denominators the same.
- The vertically and crosswise method allows you to solve problems involving any fractions.
- To convert a mixed number into a fraction, multiply the whole number by the denominator and add it to the numerator.
- The easiest way to divide fractions is to reverse the second fraction and multiply it by the first fraction.

CHAPTER

7

Decimals

In This Chapter

- Understanding the decimal system
- Adding and subtracting numbers with decimals
- Multiplying and dividing numbers with decimals

Like fractions, working with decimals is something many people find a bit difficult to handle. For any type of problem, the placement of decimals is a precise and sometimes confusing process. In this chapter, I throw some light on the concept of decimals and show you how to make it a little easier.

The Decimal System and Place Value

In the decimal number system, the position of the number determines its value—this is known as *place value.*

The principle of place value is that each place has a value 10 times the place to its right. Inversely, a position to the right is 10 times smaller than a value on the left. For example, in the number 47, the

4 is in the tens place, meaning it has a value of four 10s, while the 7 is in the ones place, meaning it has a value of seven 1s. Without place value, calculations would be extremely difficult, because place value helps us understand the meaning of a number. If people just used numbers randomly, no one would know which numbers people meant. You need place value to understand the order of numbers as well.

The decimal point is used to distinguish between whole numbers and parts of a whole. For example, 0.1 is one-tenth part of one, 0.01 is a one-hundredth part of one, and 0.001 is one-thousandth part of one.

The following table gives you a breakdown of the place values around a decimal point. As you can see, numbers get larger the farther left you go and smaller the farther right you go.

Thousands	Hundreds	Tens	Ones	.	Tenths	Hundredths	Thousandths
TH	H	T	U	.	t	H	th

Adding Decimal Numbers

When adding numbers with decimals, keep the decimal points in a vertical line. You can do the problems from left to right or right to left. In some cases, you can make the process even easier by first adding without the decimals and then putting in the decimal point at the end.

<u>Example 1</u>

Solve the problem 4.34 + 3.42.

Step 1: For this problem, ignore the decimal point and add the two numbers.

$$\begin{array}{r} 434 \\ +\,342 \\ \hline 776 \end{array}$$

Step 2: Put the decimal place back in the answer. Note that in these numbers, the decimal point is two places before the end digit. Therefore, you should put the decimal point two places before the end digit.

$$\begin{array}{r} 4.34 \\ +\,3.42 \\ \hline 7.76 \end{array}$$

Solution: The answer is 7.76.

<u>**Example 2**</u>

Solve the problem 78.3 + 2.031 + 2.3245 + 9.2.

Step 1: Set up the problem vertically, with the decimal points in alignment.

$$\begin{array}{l} \;\;78.3 \\ \;\;\;\;2.031 \\ \;\;\;\;2.3245 \\ +\;9.2 \\ \hline \end{array}$$

Step 2: Add the numbers.

$$\begin{array}{l} \;\;78.3 \\ \;\;\;\;2.031 \\ \;\;\;\;2.3245 \\ +\;\;9.2 \\ \hline \;\;91.8555 \end{array}$$

Solution: The answer is 91.8555.

QUICK TIP

You can also add zeroes to the numbers in example 2—making them 78.3000, 02.031, 02.3245, and 09.2000—so they all have the name number of digits both before and after the decimal.

<u>Example 3</u>

Solve the problem 0.0004 + 6.32 + 1.008 + 3.452.

Step 1: Set up the problem vertically, with the decimal points in alignment.

$$\begin{array}{r} 0.0004 \\ 6.32 \\ 1.008 \\ +\ 3.452 \\ \hline \end{array}$$

Step 2: Add the numbers.

$$\begin{array}{r} 0.0004 \\ 6.32 \\ 1.008 \\ +\ 3.452 \\ \hline 10.7804 \end{array}$$

Solution: The answer is 10.7804.

Subtracting Decimal Numbers

The process for subtracting numbers with decimals isn't that different from what you do when adding them. When subtracting decimal numbers, you first line up the decimal points—tens under tens, ones under ones, and so on. This makes it simpler to understand and solve the problem.

<u>Example 1</u>

Solve the problem 45 – 2.09.

Step 1: Set up the problem vertically, with the decimal points in alignment. Write 45 as 45.00 because 2.09 has two digits after the decimal points. Adding the zeroes makes it easier and safer to get the answer, because the ones are below the ones, the tens are under the tens, and so on.

$$\begin{array}{r} 45.00 \\ -\ 2.09 \\ \hline \end{array}$$

Step 2: Subtract without considering the decimal points: 4,500 – 209 = 4,291. Put the decimal point in the answer again.

```
  45.00
 - 2.09
 ------
  42.91
```

Solution: The answer is 42.91.

<u>**Example 2**</u>

Solve the problem 7.005 – 0.55.

Step 1: Set up the problem vertically, with the decimal points in alignment. Put a zero after 0.55 so it's 0.550; this makes the subtraction easier and clearer.

```
   7.005
 - 0.550
 -------
```

Step 2: Subtract without considering the decimal points: 7,005 – 0,550 = 6,455. Put the decimal point in the answer again.

```
   7.005
 - 0.550
 -------
   6.455
```

Solution: The answer is 6.455.

<u>**Example 3**</u>

Solve the problem 19.19 – 3.3.

Step 1: Set up the problem vertically, with the decimal points in alignment. Add zeroes around 3.3 so it's 03.30.

```
   19.19
 - 03.30
 -------
```

Step 2: Subtract without considering the decimal points: 1,919 – 0,330 = 1,589. Put the decimal point in the answer again.

```
  19.19
- 03.30
-------
  15.89
```

Solution: The answer is 15.89.

Multiplying Decimal Numbers

Knowing how to multiply numbers in relation to a decimal point can be frustrating. Where should the decimal go? What order do you calculate the numbers? Let me clear up any confusion for you by showing you some tips and tricks for multiplying numbers with decimals.

Multiplying by Powers of 10

Multiplying a number with a decimal by a power of 10 is very easy; it just involves moving the decimal point.

Say you have to multiply 7.86 by 10. Because 10 has one zero, you only need to move the decimal point one place to the right to get your answer: 78.6. And if you're multiplying 7.86 by 100, you move the decimal point two places to right to account for the two zeroes in 100. This gives you an answer of 786. Simple, right?

The following are examples of what happens to a number with a decimal when it's multiplied by different powers of 10. You should have no trouble calculating these in your head once you get comfortable with how you need to move the decimal.

Number	× 10	× 100	× 1,000	× 10,000
.72	7.2	72	720	7,200
.04	.4	4	40	400
5.04	50.4	504	5,040	50,400
42.03	420.3	4,203	42,030	420,300
561.321	5,613.21	56,132.1	561,321	5,613,210

Multiplying Vertically and Crosswise

It's very easy to multiply numbers with decimals using the vertical and crosswise method.

If you recall, to do the vertically and crosswise method for two-digit numbers, you first multiply the right column. You then cross-multiply the numbers, and finish by multiplying the left column.

Multiplying a three-digit number by another three-digit number requires a couple more steps in the vertically and crosswise method—vertical, crosswise, star, crosswise, and vertical. See Chapter 1 if you need a visual, and keep these two versions in your head as you work the following examples.

<u>Example 1</u>

Solve the problem 7.3 × 1.4.

Step 1: Ignoring the decimals at first and following the vertically and crosswise pattern, begin by multiplying the right column: 4 × 3 = 12. Put down 2 and carry the 1 to the next step.

$$\begin{array}{r} 7.3 \\ \times\, 1.4 \\ \hline {}_{1}2 \end{array}$$

Step 2: Cross-multiply and add: (4 × 7) + (1 × 3) = 28 + 3 = 31. Add the carryover: 31 + 1 = 32. Put down 2 and carry the 3 to the next step.

$$\begin{array}{r} 7.3 \\ \times\, 1.4 \\ \hline {}_{3}22 \end{array}$$

Step 3: Multiply the left column: 1 × 7 = 7. Add the carryover: 7 + 3 = 10. Put down 10.

$$\begin{array}{r} 7.3 \\ \times\, 1.4 \\ \hline 1022 \end{array}$$

Step 4: Put the decimal in the sum. In each of the numbers, the decimal point is one place from the right; therefore, the decimal point should be two places from the right.

$$\begin{array}{r} 7.3 \\ \times 1.4 \\ \hline 10.22 \end{array}$$

Solution: The answer is 10.22.

QUICK TIP

To put the decimal place in the correct spot, always count the number of places from the right in both the numbers and add them. Once you know the total number of decimal places, count over that many from the right and place the decimal point in the answer.

Example 2

Solve the problem 6.2 × 5.4.

Step 1: Ignoring the decimals and applying the vertically and crosswise formula, multiply the right column: 2 × 4 = 8. Put down 8.

$$\begin{array}{r} 6.2 \\ \times 5.4 \\ \hline 8 \end{array}$$

Step 2: Cross-multiply and add: (4 × 6) + (5 × 2) = 24 + 10 = 34. Put down 4 carry over the 3 to the next step.

$$\begin{array}{r} 6.2 \\ \times 5.4 \\ \hline {}_{3}48 \end{array}$$

Step 3: Multiply the left column: 6 × 5 = 30. Add the carryover: 30 + 3 = 33. Put down 33.

$$\begin{array}{r} 6.2 \\ \times 5.4 \\ \hline 3348 \end{array}$$

Step 4: Put the decimal in the sum. In each of the numbers, the decimal point is one place from the right; therefore, the decimal point should be two places from the right.

$$\begin{array}{r} 6.2 \\ \times\, 5.4 \\ \hline 33.48 \end{array}$$

Solution: The answer is 33.48.

<u>Example 3</u>

Solve the problem 3.42 × 71.5.

Step 1: Ignoring the decimals and applying the vertically and crosswise formula, multiply the right column: 2 × 5 = 10. Put down the 0 and carry over the 1.

$$\begin{array}{r} 3.42 \\ \times\, 71.5 \\ \hline {}_{1}0 \end{array}$$

Step 2: Cross-multiply the numbers in the right and middle columns and add them: (5 × 4) + (1 × 2) = 22. Add the carryover : 22 + 1 = 23. Put down 3 and carry over the 2 to the next step.

$$\begin{array}{r} 3.42 \\ \times\, 71.5 \\ \hline {}_{2}30 \end{array}$$

Step 3: Cross-multiply the numbers in the left and right columns, vertically multiply the numbers in the center, and add them: (5 × 3) + (7 × 2) + (1 × 4) = 33. Add the carryover: 33 + 2 = 35. Put down 5 and carry over the 3 to the next step.

$$\begin{array}{r} 3.42 \\ \times\, 71.5 \\ \hline {}_{3}530 \end{array}$$

Step 4: Cross-multiply the numbers in the left and center columns and add them: (1 × 3) + (7 × 4) = 31. Add the carryover: 31 + 3 = 34. Put down 4 and carry over the 3 to the next step.

$$\begin{array}{r} 3.42 \\ \times\, 71.5 \\ \hline {}_{3}4530 \end{array}$$

Step 5: Multiply the left column: 7 × 3 = 21. Add the carryover: 21 + 3 = 24. Put down 24.

$$\begin{array}{r} 3.42 \\ \times\, 71.5 \\ \hline 244530 \end{array}$$

Step 6: Put the decimal in the sum. In 3.42, the decimal point is two places from the right; in 71.5, the decimal point is one place from the right. Therefore, the decimal should be three places from the right in the answer.

$$\begin{array}{r} 3.42 \\ \times\, 71.5 \\ \hline 244.530 \end{array}$$

Solution: The answer is 244.530.

Dividing Decimal Numbers

Now that you've learned how to multiply decimal numbers, you can take on a more complex process—dividing decimal numbers. In the following sections, I give you some different scenarios for dividing decimals and ways to painlessly solve the problems.

Dividing by Powers of 10

The process for dividing by a power of 10 is exactly opposite to the process of multiplying by a power of 10. When you divide by a power of 10, instead of moving the decimal point to the right, you move it to the left to get your answer.

The number of places you move the decimal point depends on the number of zeroes in the power of 10. For example, in the problem 3.17 ÷ 10, the 10 has one zero; therefore, you move the decimal point one place to the left to get your answer: 0.317. If you were dividing by 100 instead, you'd move the decimal two places to the left, making the answer 0.0317.

The following are some examples of what happens to a number when it's divided by different powers of 10. Like multiplying decimals, you should have little trouble calculating these in your hand once you understand the placement of the decimal in relation to the number of zeroes in the power of 10.

Number	**÷ 10**	**÷ 100**	**÷ 1,000**	**÷ 10,000**
0.73	0.073	0.0073	0.00073	0.000073
8.432	0.8432	0.08432	0.008432	0.000843
657.745	65.7745	6.57745	0.657745	0.065775
6,894.942	689.4942	68.94942	6.894942	0.689494
93.05	9.305	0.9305	0.09305	0.009305
67,823.437	6,782.344	678.2344	67.82344	6.782344

Dividing a Decimal Number by a Whole Number

The fastest and easiest way to divide a decimal number by a whole number is to remove the decimal point and treat them both as whole numbers. Once you get the sum, you then add the decimal back in the same place as it was in the dividend.

Example 1

Solve the problem 9.1 ÷ 7.

Step 1: Remove the decimal and divide the two numbers.

$91 \div 7 = 13$

Step 2: Put the decimal point in your answer, in the same place as the dividend. Because the decimal point in 9.1 is one place from the right, you should put the decimal one place from the right in your answer.

Solution: The answer is 1.3.

Example 2

Solve the problem 5.26 ÷ 2.

Step 1: Remove the decimal and divide the two numbers.

526 ÷ 2 = 263

Step 2: Put the decimal point in our answer, right in the same place as it was in the dividend. Because the decimal point in 5.26 is two places from the right, you should put the decimal two places from the right in your answer.

Solution: The answer is 2.63.

SPEED BUMP

Because it's so much simpler to divide without the decimal in place, you may forget about it entirely. Remember to add back the decimal at the end!

Dividing a Decimal Number by Another Decimal Number

So what do you do when you want to divide a decimal number by another decimal number? The trick is to convert the numbers you're dividing by to whole numbers first by shifting the decimal points to the right.

Example 1

Solve the problem 567.29 ÷ 45.67.

Step 1: Because the decimal is two places to the right in each number, you shift the decimal points right two places to get the whole numbers. So 567.29 becomes 56,729 and 45.67 becomes 4,567.

Step 2: Divide the numbers.

56729 ÷ 4567 = 12.4215

Solution: The answer is 12.4215.

Example 2

Solve the problem $7.625 \div 0.923$.

Step 1: Because the decimal is three places to the right in each number, you shift the decimal points right three places to get the whole numbers. So 7.625 becomes 7,625 and 0.923 becomes 923.

Step 2: Divide the numbers.

$$7625 \div 923 = 8.2611$$

Solution: The answer is 8.2611.

Example 3

Solve the problem $52.3 \div 8.1$.

Step 1: Because the decimal places are in different places in the numbers, multiply both numbers by 10 in order to remove the decimal points.

$$52.3 \times 10 = 523$$
$$8.1 \times 10 = 81$$

Step 2: Divide the two numbers. You don't need to add the decimal back to the answer, because you multiplied both by the same base.

$$523 \div 81 = 6{,}456$$

Solution: The answer is 6.456.

The Least You Need to Know

- A decimal point is used to distinguish between whole numbers and parts of a whole.
- When adding and subtracting numbers with decimal points, ignore the decimal when doing your calculations until the very end. You can then place it back in the answer.
- To multiply two-digit or three-digit numbers with decimals, you can use the vertically and crosswise method.
- When multiplying a decimal number by a power of 10, you simply move the decimal point to the right based on the number of zeroes to get your answer; when dividing, you move the decimal point to the left.

CHAPTER

8

Percentages

In This Chapter

- Converting percentages into fractions and vice versa
- Changing percentages into decimal numbers
- Finding the percentage of a given quantity
- Getting approximate percentages
- Increasing or decreasing a quantity by a certain percentage

A percentage is a number or ratio expressed as a fraction of 100. And as you probably know, percentages are denoted by the percent sign: %. Percentages are generally used to express how big or small one quantity is when comparing it with another percentage. You probably deal with them frequently in your daily life, so why not make it easier?

In this chapter, you learn how to work with percentages more efficiently, including converting a percentage into a fraction, finding the percentage of a quantity, estimating the percentage of something, and much more.

Converting Percentages into Fractions

To convert a percentage to a fraction, all you need to do is put the number over 100 and simplify by the greatest common factor, if necessary.

Example 1

Convert 75% into a fraction.

Step 1: Turn the percentage into a fraction out of 100.

$$75\% = \frac{75}{100}$$

Step 2: Simplify the fraction by the greatest common factor. For $\frac{75}{100}$, the greatest common factor is 25. Dividing by 25 gives you the fraction $\frac{3}{4}$.

$$75\% = \frac{75}{100} = \frac{3}{4}$$

Solution: The fraction is $\frac{3}{4}$.

QUICK TIP

Don't remember what *greatest common factor* means? Let me jog your memory: it's the highest number that divides exactly into two or more numbers. For example 1, the greatest common factor is 25, because when you divide the numerator and denominator by it, you get a number without a remainder or decimal.

Example 2

Convert 40% into a fraction.

Step 1: Turn the percentage into a fraction out of 100.

$$40\% = \frac{40}{100}$$

Step 2: Simplify the fraction by the greatest common factor. For $\frac{40}{100}$, the greatest common factor is 20. Dividing by 20 gives you the fraction $\frac{2}{5}$.

$$40\% = \frac{40}{100} = \frac{2}{5}$$

Solution: The fraction is $\frac{2}{5}$.

Example 3

Convert 65% into a fraction.

Step 1: Turn the percentage into a fraction out of 100.

$$65\% = \frac{65}{100}$$

Step 2: Simplify the fraction by the greatest common factor. For $\frac{65}{100}$, the greatest common factor is 5. Dividing by 5 gives you the fraction $\frac{13}{20}$.

$$65\% = \frac{65}{100} = \frac{13}{20}$$

Solution: The fraction is $\frac{13}{20}$.

Converting Fractions into Percentages

To convert a fraction to a percentage, you do the exact opposite of what you did in the previous section—you multiply by 100.

<u>Example 1</u>

Convert $\frac{3}{4}$ into a percentage.

Multiply the fraction by 100. You can reduce by the greatest common factor (in this case, 4) before multiplying to get your answer.

$$\frac{3}{\cancel{4}_1} \times \overset{25}{\cancel{100}} = 3 \times 25 = 75\%$$

Solution: The percentage is 75%.

<u>Example 2</u>

Convert $\frac{2}{5}$ into a percentage.

Multiply the fraction by 100. You can reduce by the greatest common factor (in this case, 5) before multiplying to get your answer.

$$\frac{2}{\cancel{5}_1} \times \overset{20}{\cancel{100}} = 2 \times 20 = 40\%$$

Solution: The percentage is 40%.

<u>Example 3</u>

Convert $\frac{17}{20}$ into a percentage.

Multiply the fraction by 100. You can reduce by the greatest common factor (in this case, 20) before multiplying to get your answer.

$$\frac{17}{\cancel{20}_1} \times \overset{5}{\cancel{100}} = 17 \times 5 = 85\%$$

Solution: The percentage is 85%.

Converting Percentages into Decimals

To convert percentages into decimals, all you need to do is move the decimal two places to the left.

Example 1

Convert 45.6% into a decimal.

Because 45.6% is technically 45.6 divided by 100, simply move the decimal two places to the left.

$$\frac{45.6}{100} = 0.456$$

Solution: The decimal is 0.456.

Example 2

Convert 8.09% into a decimal.

This is the same as dividing 8.09 by 100, so all you need to do is shift the decimal two places to the left.

$$\frac{8.09}{100} = 0.0809$$

Solution: The decimal is 0.0809.

Example 3

Convert 0.674% into a decimal.

Because you're basically dividing .674 by 100, just shift the decimal point two places to the left.

$$\frac{0.674}{100} = 0.00674$$

Solution: The decimal is 0.00674.

Finding the Percentage of a Given Quantity

You can find the percentage of a given quantity by using our old friend, the vertically and crosswise method (see Chapter 2 for the process).

Two-Digit Numbers

If the percentage and quantity are two-digit numbers, you can use the two-digit version of the vertically and crosswise method.

Example 1

Find 45% of 81.

Step 1: Multiply 45 and 81 using the vertically and crosswise method. For two-digit numbers, that means you start by multiplying the right column: 5 × 1 = 5. Put down 5.

$$\frac{45}{100} \times 81 = \quad 5$$

Step 2: Multiply crosswise and add: (4 × 1) + (5 × 8) = 4 + 40 = 44. Put down 4 and carry over the 4 to the next step.

$$\frac{45}{100} \times 81 = {}_{4}45$$

Step 3: Multiply the left column: 4 × 8 = 32. Add the carryover: 32 + 4 = 36. Put down 36.

$$\frac{45}{100} \times 81 = 3645$$

Step 4: Put the decimal in the answer. Because you're dividing 3,645 by 100, put the decimal two places to the left.

$$\frac{45}{100} \times 81 = \frac{3645}{100} = 36.45$$

Solution: The answer is 36.45.

Example 2

Find 73% of 98.

Step 1: Multiply 73 and 98 using the vertically and crosswise method. Begin by multiplying the right column: $3 \times 8 = 24$. Put down 4 and carry over the 2 to the next step.

$$\frac{73}{100} \times 98 = {}_{2}4$$

Step 2: Multiply crosswise and add: $(7 \times 8) + (3 \times 9) = 56 + 27 = 83$. Add the carryover: $83 + 2 = 85$. Put down 5 and carry over the 8 to the next step.

$$\frac{73}{100} \times 98 = {}_{8}54$$

Step 3: Multiply the left column: $7 \times 9 = 63$. Add the carryover: $63 + 8 = 71$. Put down 71.

$$\frac{73}{100} \times 98 = 7154$$

Step 4: Put the decimal in the answer. Because you're dividing 7,154 by 100, put the decimal two places to the left.

73 -- 7154 _. _.

Solution: The answer is 71.54.

Example 3

Find 23% of 67.

Step 1: Multiply 23 and 67 using the vertically and crosswise method. Start with the right column: $3 \times 7 = 21$. Put down 1 and carry over the 2 to the next step.

$$\frac{23}{100} \times 67 = {}_{2}1$$

Step 2: Multiply crosswise and add: (3 × 6) + (2 × 7) = 18 + 14 = 32. Add the carryover: 32 + 2 = 34. Put down 4 and carry over the 3 to the next step.

$$\frac{23}{100} \times 67 = \quad {}_{3}41$$

Step 3: Multiply the left column: 2 × 6 = 12. Add the carryover: 12 + 3 = 15. Put down 15.

$$\frac{23}{100} \times 67 = 1541$$

Step 4: Put the decimal in the answer. Because you're dividing 1,541 by 100, put the decimal two places to the left.

$$\frac{23}{100} \times 67 = \frac{1541}{100} = 15.41$$

Solution: The answer is 15.41.

Three-Digit Numbers

If the percentage and quantity are three-digit numbers, you can use the three-digit version of the vertically and crosswise method.

QUICK TIP

If you'd like a visual of the multiplication steps for the three-digit version of the vertically and crosswise method, flip back to Chapter 2.

Example 1

Find 14.2% of 682.

Step 1: Ignoring the decimal point, multiply 142 and 682 using the vertically and crosswise method. Begin by multiplying the right column: 2 × 2 = 4. Put down 4.

$$\frac{14.2}{100} \times 682 = \quad 4$$

Step 2: Multiply the right and middle columns crosswise and add: (2 × 4) + (8 × 2) = 8 + 16 = 24. Put down 4 and carry over the 2 to the next step.

$$\frac{14.2}{100} \times 682 = {}_{2}44$$

Step 3: Multiply in the star pattern—the bottom-right and top-left, the top and bottom middle, and the bottom-left and top-right numbers—and add: (2 × 1) + (8 × 4) + (6 × 2) = 2 + 32 + 12 = 46. Add the carryover: 46 + 2 = 48. Put down 8 and carry over the 4 to the next step.

$$\frac{14.2}{100} \times 682 = {}_{4}844$$

Step 4: Multiply the right and middle columns crosswise and add: (6 × 4) + (8 × 1) = 24 + 8 = 32. Add the carryover: 32 + 4 = 36. Put down 6 and carry over the 3 to the next step.

$$\frac{14.2}{100} \times 682 = {}_{3}6844$$

Step 5: Multiply left column: 6 × 1 = 6. Add the carryover: 6 + 3 = 9. Put down 9.

$$\frac{14.2}{100} \times 682 = 96844$$

Step 6: Put the decimal in the answer. Because the percentage had a decimal one to the left, place it there so 96,844 becomes 9,684.4. Now you have to divide 9,684.4 by 100, which simply means putting the decimal two places to the left.

$$\frac{14.2}{100} \times 682 = \frac{9684.4}{100} = 96.844$$

Solution: The answer is 96.844.

Example 2

Find 71.1% of 475.

Step 1: Ignoring the decimal point, multiply 711 and 475 using the vertically and crosswise method. Begin by multiplying the right column: 5 × 1 = 5. Put down 5.

$$\frac{71.1}{100} \times 475 = \quad 5$$

Step 2: Multiply the right and middle columns crosswise and add: (5 × 1) + (7 × 1) = 5 + 7 = 12. Put down 2 and carry over the 1 to the next step.

$$\frac{71.1}{100} \times 475 = \quad {}_{1}25$$

Step 3: Multiply in the star pattern—the bottom-right and top-left, the top and bottom middle, and the bottom-left and top-right numbers—and add: (5 × 7) + (7 × 1) + (4 × 1) = 35 + 7 + 4 = 46. Add the carryover: 46 + 1 = 47. Put down 7 and carry over the 4 to the next step.

$$\frac{71.1}{100} \times 475 = \quad {}_{4}725$$

Step 4: Multiply the right and middle columns crosswise and add: (4 × 1) + (7 × 7) = 4 + 49 = 53. Add the carryover: 53 + 4 = 57. Put down 7 and carry over the 5.

$$\frac{71.1}{100} \times 475 = \quad {}_{5}7725$$

Step 5: Multiply the right column: 4 × 7 = 28. Add the carryover: 28 + 5 = 33. Put down 33.

$$\frac{71.1}{100} \times 475 = 337725$$

Step 6: Put the decimal in the answer. Because the percentage had a decimal one to the left, place it there so 337,725 becomes 33,772.5. Now you have to divide 33,772.5 by 100, which simply means putting the decimal two places to the left.

$$\frac{71.1}{100} \times 475 = \frac{33772.5}{100} = 337.725$$

Solution: The answer is 337.725.

Example 3

Find 87.2% of 584.

Step 1: Ignoring the decimal point, multiply 872 and 584 using the vertically and crosswise method. Begin by multiplying the right column: 4 × 2 = 8. Put down 8.

$$\frac{87.2}{100} \times 584 = \qquad 8$$

Step 2: Multiply the right and middle columns crosswise and add: (4 × 7) + (8 × 2) = 28 + 16 = 44. Put down 4 and carry over the 4 to the next step.

$$\frac{87.2}{100} \times 584 = \qquad {}_{4}48$$

Step 3: Multiply in the star pattern—the bottom-right and top-left, the top and bottom middle, and the bottom-left and top-right numbers—and add: (5 × 2) + (8 × 7) + (4 × 8) = 10 + 56 + 32 = 98. Add the carryover: 98 + 4 = 102. Put down 2 and carry over the 10 to the next step.

$$\frac{87.2}{100} \times 584 = \qquad {}_{10}248$$

Step 4: Multiply the right and middle columns crosswise and add: (8 × 8) + (7 × 5) = 64 + 35 = 99. Add the carryover: 99 + 10 = 109. Put down 9 and carry over the 10 to the next step.

$$\frac{87.2}{100} \times 584 = {}_{10}9248$$

Step 5: Multiply the right column: 5 × 8 = 40. Add the carryover: 40 + 10 = 50. Put down 50.

$$\frac{87.2}{100} \times 584 = 509248$$

Step 6: Put the decimal in the answer. Because the percentage had a decimal one to the left, place it there so 509,248 becomes 50,924.8. Now you have to divide 50,924.8 by 100, which simply means putting the decimal two places to the left.

$$\frac{87.2}{100} \times 584 = \frac{50924.8}{100} = 509.248$$

Solution: The answer is 509.248.

Expressing One Quantity as a Percentage of Another

For this scenario, all you have to do is divide the numbers and then multiply the answer by 100 to get the percentage.

Example 1

Find 75 as a percentage of 250.

Step 1: Divide 75 by 250.

$75 \div 250 = .3$

Step 2: Multiply the number by 100. If you recall, you simply need to move the decimal two places to the right to get the number.

$.3 \times 100 = 30$

Solution: The answer is 30%.

Example 2

Find 82 as a percentage of 450.

Step 1: Divide 82 by 450.

$82 \div 450 = .1822$

Step 2: Multiply the number by 100. If you recall, you simply need to move the decimal two places to the right to get the number.

$.1822 \times 100 = 18.22$

Solution: The answer is 18.22%.

<u>Example 3</u>

Find 35 as a percentage of 890.

Step 1: Divide 35 by 890.

$35 \div 890 = .03932$

Step 2: Multiply the number by 100. If you recall, you simply need to move the decimal two places to the right to get the number.

$.03932 \times 100 = 3.932$

Solution: The answer is 3.932%.

Approximating Percentages

Sometimes, you may need to get an idea or estimate of how much one number is a percentage of the other. Without the aid of a calculator, you can use the percentages you do know for the larger number and build off that information to get the answer.

<u>Example 1</u>

Estimate 42 as a percentage of 800.

Step 1: Start with the simplest percentage that will reduce the larger number to two digits. For 800, you can use 10%, which gives you 80.

Step 2: Build off that percentage until you get the parts of your answer. Half of 10% is 5%, and half of 80 is 40; that gives you one piece of your answer, because it's close to 42. To get the percentage for 2, start with 1% of 800, which is 8. The 2 is one fourth of 8, so the percentage has to be one fourth of 1%, or 0.25%.

$10\% \text{ of } 800 = 80$
$5\% \text{ of } 800 = 40$
$1\% \text{ of } 800 = 8$
$0.25\% \text{ of } 800 = 2$

Step 3: Add the percentages you found for 40 and 2.

5% + 0.25% = 5.25%.

Solution: The answer is 5.25%.

Example 2

Estimate 45 as a percentage of 650.

Step 1: Start with the simplest percentage that will reduce the larger number to two digits. For 650, you can use 10%, which gives you 65.

Step 2: Build off that percentage until you get the parts of your answer. Half of 10% is 5%, and half of 65 is 32.5; that gives you one piece of your answer, because it's close to 45. To get the percentage for the rest of the answer, which is a little over 12, start with 1% of 650, which is 6.5.

5% of 650 = 32.5

1% of 650 = 6.5

Step 3: Add the percentages you found for 32.5 and 6.5; in case, add the 6.5 twice, because the remainder above was over 12. The number equal to this percentage is 45.5, slightly higher than 45, so the percentage is a little less than what you get when you add.

5% + 1% + 1% = 7%

Solution: The answer is a little less than 7%.

Example 3

Estimate 73 as a percentage of 3,568.

Step 1: Start with the simplest percentage that will reduce the larger number to two digits. For 3,568, you can use 1%, which gives you 35.68.

Step 2: Build off that percentage until you get the parts of your answer. The double of 1% is 2%, and the double of 35.68 is 71.36; this is extremely close, so you can use this percentage as your answer. Because 71.36 is less than 73, that means your percentage is a little higher than 2%.

1% of 3568 = 35.68

2% of 3568 = 71.36

Solution: The answer is a little more than 2%.

QUICK TIP

I used percentages like 10% and 1% as starting points in the examples, because those involve simply moving the decimal to find the value. If you still have trouble with knowing where to move the decimal, feel free to revisit Chapter 7.

Percentage Increase or Decrease

To increase or decrease a number by a certain percentage, you utilize the vertically and crosswise method yet again. The following sections break down how to perform the process for each type.

Percentage Increase

To find the solution for a number increased by a percentage, using the vertically and crosswise method, multiply the number by the decimal version of the percentage plus 1. You add the 1 because you're finding how much more it is than the number.

Example 1

Increase 673 by 23%.

Step 1: The decimal version of 23% is .23; add 1 to that, so it becomes 1.23. Now, ignoring the decimal for the moment, multiply 673 and 1.23. Start with the right column: 3 × 3 = 9. Put down 9.

$$\begin{array}{r} 673 \\ \times\, 1.23 \\ \hline 9 \end{array}$$

Step 2: Multiply the right and middle columns crosswise and add: (3 × 7) + (3 × 2) = 21 + 6 = 27. Put down 7 and carry over the 2 to the next step.

$$\begin{array}{r} 673 \\ \times 1.23 \\ \hline {}_{2}79 \end{array}$$

Step 3: Multiply in the star pattern—the bottom-right and top-left, the top and bottom middle, and the bottom-left and top-right numbers—and add: (3 × 6) + (7 × 2) + (3 × 1) = 18 + 14 + 3 = 35. Add the carryover: 35 + 2 = 37. Put down 7 and carry over the 3 to the next step.

$$\begin{array}{r} 673 \\ \times 1.23 \\ \hline {}_{3}779 \end{array}$$

Step 4: Multiply the left and middle columns crosswise and add: (1 × 7) + (2 × 6) = 7 + 12 = 19. Add the carryover: 19 + 3 = 22. Put down 2 and carry over the 2 to the next step.

$$\begin{array}{r} 673 \\ \times 1.23 \\ \hline {}_{2}2779 \end{array}$$

Step 5: Multiply the left column: 1 × 6 = 6. Add the carryover: 6 + 2 = 8. Put down 8. Now add the decimal point. The decimal in 1.23 is two places to the left, so for 82,779, put the decimal two places to the left.

$$\begin{array}{r} 673 \\ \times 1.23 \\ \hline 827.79 \end{array}$$

Solution: The answer is 827.79.

Example 2

Increase 450 by 34%.

Step 1: The decimal version of 34% is .34; add 1 to that, so it becomes 1.34. Now, ignoring the decimal for the moment, multiply 450 and 1.34. Start with the right column: $4 \times 0 = 0$. Put down 0.

$$\begin{array}{r} 450 \\ \times\, 1.34 \\ \hline 0 \end{array}$$

Step 2: Multiply the right and middle columns crosswise and add: $(4 \times 5) + (3 \times 0) = 20 + 0 = 20$. Put down 0 and carry over the 2 to the next step.

$$\begin{array}{r} 450 \\ \times\, 1.34 \\ \hline {}_{2}00 \end{array}$$

Step 3: Multiply in the star pattern—the bottom-right and top-left, the top and bottom middle, and the bottom-left and top-right numbers—and add: $(4 \times 4) + (1 \times 0) + (3 \times 5) = 16 + 0 + 15 = 31$. Add the carryover: $31 + 2 = 33$. Put down 3 and carry over the 3 to the next step.

$$\begin{array}{r} 450 \\ \times\, 1.34 \\ \hline {}_{3}300 \end{array}$$

Step 4: Multiply the left and middle columns crosswise and add: $(3 \times 4) + (1 \times 5) = 12 + 5 = 17$. Add the carryover: $17 + 3 = 20$. Put down 0 and carry over the 2 to the next step.

$$\begin{array}{r} 450 \\ \times\, 1.34 \\ \hline {}_{2}0300 \end{array}$$

Step 5: Multiply the right column: $1 \times 4 = 4$. Add the carryover: $4 + 2 = 6$. Now add the decimal point. The decimal in 1.34 is two places to the left, so for 60,300, put the decimal two places to the left.

$$\begin{array}{r} 450 \\ \times\, 1.34 \\ \hline 603.00 \end{array}$$

Solution: The answer is 603.00.

Percentage Decrease

Like you did for percentage increase, you use the vertically and crosswise method to get your answer. However, this time, you multiply the number by the decimal value of the percentage *minus* 1, because you're finding how much less it is than the number.

<u>Example 1</u>

Decrease 500 by 25%.

Step 1: The decimal version of 25% is .25; subtract 1 from that, so it becomes .75. Now, ignoring the decimal for the moment, multiply 500 and 0.75. Start with the right column: $5 \times 0 = 0$. Put down 0.

$$\begin{array}{r} 500 \\ \times\, 0.75 \\ \hline 0 \end{array}$$

Step 2: Multiply the right and middle columns crosswise and add: $(5 \times 0) + (7 \times 0) = 0 + 0 = 0$. Put down 0.

$$\begin{array}{r} 500 \\ \times\, 0.75 \\ \hline 00 \end{array}$$

Step 3: Multiply in the star pattern—the bottom-right and top-left, the top and bottom middle, and the bottom-left and top-right numbers—and add: $(5 \times 5) + (0 \times 0) + (7 \times 0) = 25 + 0 + 0 = 25$. Put down 5 and carry over the 2 to the next step.

$$\begin{array}{r} 500 \\ \times\, 0.75 \\ \hline {}_{2}500 \end{array}$$

Step 4: Multiply the left and middle columns crosswise and add: $(7 \times 5) + (0 \times 0) = 35 + 0 = 35$. Add the carryover: $35 + 2 = 37$. Put down 7 and carry over the 3 to the next step.

$$\begin{array}{r} 500 \\ \times\, 0.75 \\ \hline {}_{3}7500 \end{array}$$

Step 5: Multiply the left column: $0 \times 5 = 0$. Add the carryover: $0 + 3 = 3$. Put down 3. Now add the decimal point. The decimal in .75 is two places to the left, so for 37,500, put the decimal two places to the left.

$$\begin{array}{r} 500 \\ \times\, 0.75 \\ \hline 375.00 \end{array}$$

Solution: The answer is 375.00.

QUICK TIP

In the first step of the process, rather than subtracting the decimal version of the percentage from 1, you can just figure out what percentage added to the one in the problem equals 100%. For example 1, you probably know 75% added to 25% equals 100%, so feel free to use that shortcut to speed up your calculations.

Example 2

Decrease 878 by 62%.

Step 1: The decimal version of 62% is .62; subtract 1 from that, so it becomes .38. Now, ignoring the decimal for the moment, multiply 878 and 0.38. Start with the right column: 8 × 8 = 64. Put down 4 and carry over the 6 to the next step.

$$\begin{array}{r} 878 \\ \times\, 0.38 \\ \hline {}_{6}4 \end{array}$$

Step 2: Multiply the right and middle columns crosswise and add: (8 × 7) + (3 × 8) = 56 + 24 = 80. Add the carryover: 80 + 6 = 86. Put down 6 and carry over the 8 to the next step.

$$\begin{array}{r} 878 \\ \times\, 0.38 \\ \hline {}_{8}64 \end{array}$$

Step 3: Multiply in the star pattern—the bottom-right and top-left, the top and bottom middle, and the bottom-left and top-right numbers—and add: (8 × 8) + (0 × 8) + (3 × 7) = 64 + 0 + 21 = 85. Add the carryover: 85 + 8 = 93. Put down 3 and carry over the 9 to the next step.

$$\begin{array}{r} 878 \\ \times\, 0.38 \\ \hline {}_{9}364 \end{array}$$

Step 4: Multiply the left and middle columns crosswise and add: (3 × 8) + (0 × 7) = 24 + 0 = 24. Add the carryover: 24 + 9 = 33. Put down 3 and carry over the 3 to the next step.

$$\begin{array}{r} 878 \\ \times 0.38 \\ \hline {}_{3}3364 \end{array}$$

Step 5: Multiply the left column: 0 × 8 = 0. Add the carryover: 0 + 3 = 3. Put down 3. Now add the decimal point. The decimal in .38 is two places to the left, so for 33,364, put the decimal two places to the left.

$$\begin{array}{r} 878 \\ \times 0.38 \\ \hline 333.64 \end{array}$$

Solution: The answer is 333.64.

Example 3

Decrease 345 by 18%.

Step 1: The decimal version of 18% is .18; subtract 1 from that, so it becomes .82. Now, ignoring the decimal for the moment, multiply 878 and 0.82. Start with the right column: 2 × 5 = 10. Put down 0 and carry over the 1 to the next step.

$$\begin{array}{r} 345 \\ \times 0.82 \\ \hline {}_{1}0 \end{array}$$

Step 2: Multiply the right and middle columns crosswise and add: (2 × 4) + (8 × 5) = 8 + 40 = 48. Add the carryover: 48 + 1 = 49. Put down 9 and carry over the 4 to the next step.

$$\begin{array}{r} 345 \\ \times 0.82 \\ \hline {}_{4}90 \end{array}$$

Step 3: Multiply in the star pattern—the bottom-right and top-left, the top and bottom middle, and the bottom-left and top-right numbers—and add: (2 × 3) + (0 × 5) + (8 × 4) = 6 + 0 + 32 = 38. Add the carryover: 38 + 4 = 42. Put down 2 and carry over the 4 to the next step.

$$\begin{array}{r} 345 \\ \times\, 0.82 \\ \hline {}_4 290 \end{array}$$

Step 4: Multiply the left and middle columns crosswise and add: (8 × 3) + (0 × 4) = 24 + 0 = 24. Add the carryover: 24 + 4 = 28. Put down 8 and carry over the 2 to the next step.

$$\begin{array}{r} 345 \\ \times\, 0.82 \\ \hline {}_2 8290 \end{array}$$

Step 5: Multiply the left column: 0 × 3 = 0. Add the carryover: 0 + 2 = 2. Put down 2. Now add the decimal point. The decimal in .82 is two places to the left, so for 28,290, put the decimal two places to the left.

$$\begin{array}{r} 345 \\ \times\, 0.82 \\ \hline 282.90 \end{array}$$

Solution: The answer is 282.90.

The Least You Need to Know

- Put the percentage over 100 and simplify by the greatest common factor in order to convert a fraction to a percentage.
- When converting fractions to percentages, multiply by 100 and simplify as necessary.
- You only have to move the decimal two places to the left to change a percentage into a decimal number.
- To find the percentage of a certain quantity, multiply the quantity and the percentage using the vertically and crosswise method.
- You can easily find the approximate percentage for a number by breaking down the number into the most recognizable percentages.
- To find the percentage increase or decrease of a number, multiply the number by the decimal version of the percentage plus or minus 1.

CHAPTER

9

Divisibility

In This Chapter

- Divisibility rules for numbers under 10
- Using the osculation method to determine divisibility
- The difference between positive and negative osculators

Finding out what numbers divide into another number can be a tedious process. You may spend a lot of time just inputting numbers into the calculator. However, it doesn't have to be that difficult!

In this chapter, I take you through some common divisibility rules you can apply to your calculations. I also teach you about a method called *osculation,* which can make determining what a number's divisible by much simpler.

Divisibility Rules

If you have a divisor that's less than 10, it's very easy to find out whether a number can be divided by it.

A number is divisible by 2 if it's even, meaning the number has to end in 2, 4, 6, 8, or 0.

A number can be divided by 3 if, when you add the digits in the number, that total is divisible by 3. For example, the number 345 is divisible by 3 because its digits add to 12 (3 + 4 + 5), which is a multiple of 3.

A number can be divided by 4 if the last two digits of the number are divisible by 4. For example, in the number 8,732, the last two digits are divisible by 4, meaning the whole number can be divided by 4.

For divisibility by 5, the end digit of the number has to be a 5 or 0. It's that simple!

Checking for divisibility by 6 is a little more complicated. Because 6 is divisible by 2 and 3, a number must pass the divisibility tests for both of those. For example, the number 12 is divisible by 6 because it's even (applicable to the rules for 2) and its digits add to a number that can be divided by 3 (applicable to the rules for 3).

A number can be divided by 8 if the last three digits are divisible by 8. For example, in the number 337,312, the 312 is divisible by 8, which means the whole number can be divided by 8.

For 9, if the sum of the digits add up to 9 or a multiple of 9, you can say the number is divisible by it.

And here's another easy rule—a number is divisible by 10 if it ends in a zero.

But what about 7? That's where the following method comes in.

The Osculation Method

The osculation method can help you find the divisibility rules for numbers over 10, in particular prime numbers. You apply the osculation method using an osculator. An osculator is a number you can derive from the divisor to find out whether another number is divisible by it. A number has both a positive and negative osculator. Sounds complicated, doesn't it? Let me carefully walk you through how to find and apply each type.

Positive Osculation

If your divisor ends in 9, the positive osculator is one more than the digit before the 9. For example, for the number 29, the osculator is 3, because the digit before the 9 is 2, and one more than that is 3. Likewise, in the number 79, the positive osculator is 8, because one more than the previous digit before the 9 is 8.

But how do you find the positive osculator for numbers that don't end in 9? Depending on the last digit, you multiply so the number ends in 9:

- If a number ends in 1, multiply it by 9.
- If a number ends in 3, multiply it by 3.
- If a number ends in 7, multiply it by 7.

The result of the multiplication gives you a number that ends in 9, which you can use to find the osculator.

For example, if you need to find the osculator for 13, you simply multiply 13 by 3 to get 39. The osculator is one more than the digit before 9; in this case, it's 4. So the osculator for the number 13 is 4.

QUICK TIP

Positive osculators are typically used for numbers ending in 3 and 9, because these values are smaller than the negative osculators (which I'll discuss in the next section). For example, the positive osculator for 19 is 2, while the negative osculator is 17. In this case, using the 2 instead of the 17 for calculations is much simpler and can help you avoid errors.

The result of the osculation process, which I walk you through in the examples, should either be the divisor itself or a multiple of the divisor. That's when you know the dividend is completely divisible by the divisor. If a number starts to repeat, you can say the number is divisible. Another result that indicates the dividend is divisible by the divisor is when the result of the osculation process is zero.

The following examples show you different scenarios of the osculation process using positive osculators.

Example 1

Find out if 112 is divisible by 7.

Step 1: Find the positive osculator for 7. Because the number is 7, multiply by 7: 7 × 7 = 49. The positive osculator is one more than the digit before the 9, so the positive osculator for 7 is 5.

Step 2: Osculate 112 with 5.

$11 + (2 \times 5) = 21$

$2 + (1 \times 5) = 7$

Solution: You end up at 7, so 112 is divisible by 7.

Example 2

Find out if 1,035 is divisible by 23.

Step 1: Find the positive osculator for 23. Because the number ends in 3, multiply by 3: 23 × 3 = 69. The positive osculator is one more than the digit before 9, so the positive osculator for 23 is 7.

Step 2: Osculate 1,035 with 7.

$103 + (5 \times 7) = 138$

$13 + (8 \times 7) = 69$

$6 + (9 \times 7) = 69$

Solution: 69, a multiple of 23, is repeating, so 1,035 is divisible by 23.

Example 3

Find out if 6,308 is divisible by 38.

Step 1: 38 is a composite number made up of 2 and 19, so test for divisibility by both of those numbers. Because 6,308 is even, it's divisible by 2. To find out if it's divisible by 19, find the positive osculator for 19, which is 2.

Step 2: Osculate 6,308 with 2.

$630 + (8 \times 2) = 646$

$64 + (6 \times 2) = 76$

$7 + (6 \times 2) = 19$

Solution: 2 and 19, multiples of 38, go into 6,308, so it is divisible by 38.

Example 4

Find out if 334,455 is divisible by 39.

Step 1: Find the positive osculator for 39. Because the number ends in 9, the osculator is only one more than the digit before the 9. So for 39, the osculator is 4.

Step 2: Osculate 334,455 with 4.

$33445 + (5 \times 4) = 33465$

$3346 + (5 \times 4) = 3366$

$336 + (6 \times 4) = 360$

$36 + 0 = 36$

Solution: 36 is below the divisor, so 334,455 is not divisible by 39.

Negative Osculation

If you have a divisor that ends in 1, the negative osculator is simply the digit before the 1. For example, if the number is 51, the negative osculator is 5—you simply drop the 1. The same is true if the divisor is 81; you just drop the 1 at the end to get the negative osculator: 8.

What if the divisor doesn't end in 1? Depending on the last digit, you can find the negative osculator by multiplying:

- If a number ends in 3, multiply it by 7.
- If a number ends in 7, multiply it by 3.
- If a number ends in 9, multiply it by 9.

The result of the multiplication gives you a number that ends in 1; you can then drop the 1 to have your negative osculator.

For example, to find the negative osculator of 7, you multiply it by 3, which gives you 21. You then simply drop the 1 to get your negative osculator; for 7, it's 2.

QUICK TIP

Negative osculators are typically used for numbers ending in 1 and 7, because these values are smaller than their positive osculator versions. For example, for 81, the positive osculator is 73 and the negative osculator is 8. It's obviously much easier to use 8 than 73 for calculations!

You can say that a dividend is divisible by the divisor even when the result of the osculation process is zero, the divisor itself, or a repetition of a previous result.

The following examples show you different scenarios of the osculation process using negative osculators.

<u>Example 1</u>

Find out if 6,603 divisible by 31.

Step 1: Find the negative osculator for 31. Because the number ends in 1, simply drop the 1 to get the negative osculator, which is 3.

Step 2: Osculate 6,603 with 3.

$660 - (3 \times 3) = 651$

$65 - (1 \times 3) = 62$

$6 - (2 \times 3) = 0$

Solution: You end up at 0, which means 6,603 is divisible by 31.

<u>Example 2</u>

Find out if 11,234 is divisible by 41.

Step 1: Find the negative osculator for 41. Because the number ends in 1, simply drop the 1 to get the negative osculator, which is 4.

Step 2: Osculate 11,234 with 4.

$1123 - (4 \times 4) = 1107$

$110 - (7 \times 4) = 82$

$8 - (2 \times 4) = 0$

Solution: You end up at 0, which indicates 11,234 is divisible by 41.

Example 3

Find out if 2,275 is divisible by 7.

Step 1: Find the negative osculator for 7. Because the number is 7, multiply by 3: $7 \times 3 = 21$. The negative osculator is the digit before the 1, so for 7, it's 2.

Step 2: Osculate 2,275 with 2.

$227 - (5 \times 2) = 217$

$21 - (7 \times 2) = 7$

Solution: You end up at 7, which means 2,275 is divisible by 7.

Example 4

Find out if 464,411 is divisible by 71.

Step 1: Find the negative osculator for 71. Because the number ends in 1, drop the 1 to get the osculator. For 71, the negative osculator is 7.

Step 2: Osculate 464,411 by 7.

$46441 - (1 \times 7) = 46434$

$4643 - (4 \times 7) = 4615$

$461 - (5 \times 7) = 426$

$42 - (6 \times 7) = 0$

Solution: You end up at 0, which indicates 464,411 is divisible by 71.

QUICK TIP

The sum of the positive and negative osculators equal the divisor. Therefore, if you already know one of the osculators, you can subtract that value from the divisor to get the other osculator. For example, the negative osculator for 61 is 6. You can then subtract 6 from 61 to get the positive osculator, which is 55.

The Least You Need to Know

- If a divisor is under 10, you can follow its divisibility rules to determine whether it can divide into a number.
- For a divisor that ends in 9, the positive osculator is one more than the digit before the 9.
- For a divisor that ends in 1, the negative osculator is the digit before the 1.

CHAPTER

10

Squared Numbers

In This Chapter

- Finding the value of a squared number that ends in 5
- Learning how to find the square of a number near 50
- Using the duplex to find the value of two-, three-, and four-digit numbers
- Figuring the sums of squares

Squared numbers are only too easy to type into a calculator, but if you need to do them in your head, it can be a bit tricky. However, there are ways you can find the values that allow you to go beyond the calculator and really understand squared numbers.

In this chapter, you learn processes for finding the value of squares and for adding squared numbers.

Finding Squares of Numbers Ending in 5

It's very simple to find the square of numbers that end with 5. To get the first part of the answer, you apply the method called *by one more than the one before.* What does this mean? It means you simply find out what's one more than the number before 5 and multiply those two numbers together. You then get the second part of your answer by squaring 5, or multiplying 5 by itself. Check out the following examples to see what I mean.

Example 1

Find the value of 35^2.

Step 1: To get the first part of the answer, apply the "by one more than the one before" method. What is one more than the first digit? One more than 3 is 4. Multiply the two numbers together.

$3 \times 4 = 12$

Step 2: To get the second part of the answer, square 5.

$5 \times 5 = 25$

Solution: The answer is 1,225.

Example 2

Find the value of 75^2.

Step 1: To get the first part, find out what's one more than the first digit and multiply. One more than 7 is 8, so multiply them together.

$7 \times 8 = 56$

Step 2: To get the second part of the answer, square 5.

$5 \times 5 = 25$

Solution: The answer is 5,625.

Keep in mind that this technique is only for finding out the squares of numbers that end with 5. Don't try to use it with squares of other numbers!

<u>Example 3</u>

Find the value of 115^2.

Step 1: To get the first part, find out what's one more than the first two digits and multiply. One more than 11 is 12, so multiply them together.

$11 \times 12 = 132$

Step 2: To get the second part of the answer, square 5.

$5 \times 5 = 25$

Solution: The answer is 13,225.

Finding Squares of Numbers Near 50

Finding the value of squared numbers near 50 is easy and fun. Like the method for numbers ending in 5, you only need to complete two steps to get the answer. Both steps involve using the excess or deficit—or how much more or less the number is than 50—to get the answer.

Numbers Above 50

If the number's above 50, you get the first part by adding the excess to 25, and you get the second part by squaring the excess. Try this process out with the following examples.

<u>Example 1</u>

Find the value of 54^2.

Step 1: To get the first part of the answer, find out how much more the number is than 50 and add the excess to 25. In this case, 54 is 4 more than 50, so add that to 25.

$25 + 4 = 29$

Step 2: To get the second part of the answer, square the excess.

$4 \times 4 = 16$

Solution: The answer is 2,916.

Example 2

Find the value of 52^2.

Step 1: To get the first part, find out how much more the number is than 50 and add the excess to 25. In this case, 52 is 2 more than 50, so add that to 25.

$25 + 2 = 27$

Step 2: To get the second part, square the excess. The second part must be a double-digit number, so put a zero before this number.

$2 \times 2 = 4$ or 04

Solution: The answer is 2,704.

Example 3

Find the value of 61^2.

Step 1: To get the first part, find out how much more the number is than 50 and add the excess to 25. In this case, 61 is 11 more than 50, so add that to 25.

$25 + 11 = 36$

Step 2: To get the second part, square the excess. Because you get a three-digit number, add the first 1 to 36. That makes the two parts of the answer 37 and 21.

$11 \times 11 = 121$

$1 + 36 = 37$

Solution: The answer is 3,721.

Numbers Below 50

If the number's below 50, instead of adding the difference to 25, you subtract it from 25. To get the second part of the answer, you do as you did before—square the difference between the number and 50. Let me show you how this works.

Example 1

Find the value of 46^2.

Step 1: To get the first part of the answer, find out how much less the number is than 50 and subtract the deficit from 25. In this case, 46 is 4 less than 50, so subtract that from 25.

$25 - 4 = 21$

Step 2: To get the second part of the answer, square the deficit.

$4 \times 4 = 16$

Solution: The answer is 2,116.

Example 2

Find the value of 42^2.

Step 1: To get the first part, find out how much less the number is than 50 and subtract the deficit from 25. In this case, 42 is 8 less than 50, so subtract that from 25.

$25 - 8 = 17$

Step 2: To get the second part, square the deficit.

$8 \times 8 = 64$

Solution: The answer is 1,764.

Example 3

Find the value of 38^2.

Step 1: To get the first part, find out how much less the number is than 50 and subtract the deficit from 25. In this case, 38 is 12 less than 50, so subtract that from 25.

$25 - 12 = 13$

Step 2: To get the second part, square the deficit. Because you get a three-digit number, add the first 1 to 13. That makes the two parts of the answer 14 and 44.

$12 \times 12 = 144$

$13 + 1 = 14$

Solution: The answer is 1,444.

Using the Duplex to Find the Value of a Squared Number

To find out the squares of any given number, you have to learn a concept called the *duplex*. The duplex has a different meaning based on the number of digits of what's being squared.

If the number being squared is just one digit, you simply use the formula a^2, where a is the number. So, for example, the duplex of 7 is 49, because $7^2 = 49$.

If the squared number has two digits, you use the formula $2ab$, where a is the first digit and b is the second digit. Say you want to know the duplex of 81; you multiply the individual digits with 2 and get your answer: $2 \times 8 \times 1 = 16$. So 81 has a duplex of 16. You can add variables based on the number of digits.

If the squared number has three digits, you go with the formula $b^2 + 2ac$, where a is the first digit, b is the second digit, and c is the third digit. Say you want to find the duplex of 372. Here's how you'd plug it in: $7^2 + (2 \times 3 \times 2) = 49 + 12 = 61$. As you can see, the duplex of 372 is 61.

If the number being squared has four digits, you use the formula $2ad + 2bc$. For example, here's how you get the duplex of 7,351: $(2 \times 7 \times 1) + (2 \times 3 \times 5) = 14 + 30 = 44$.

But how does this apply to finding the value of squares? You do the duplex for every combination in a number and combine them to get the answer. The following sections show you how this is done.

Two-Digit Squares

To find the square of any two-digit number, you find three duplexes: one for the first digit (a^2), one for the digit as a whole ($2ab$), and one for the second digit (a^2). You then combine them to get your answer.

Example 1

Find the value of 57^2.

Step 1: Find the duplexes for 5, 57, and 7 and put the three values together with slashes separating them; the slashes represent columns.

Duplex of 5: $5^2 = 25$
Duplex of 57: $2 \times 5 \times 7 = 70$
Duplex of 7: $7^2 = 49$
25/70/49

Step 2: Add the three values from right to left. You can only have one digit in every column except the last, so put down 9 for this column and carry over the 4 to the next column.

25/70/49
9

Step 3: Add the carryover to 70: 70 + 4 = 74. Again, you can only have one digit in every column except the last, so put down 4 and carry over the 7 to the next column.

25/70/49
49

Step 4: Add the carryover to 25: 25 + 7 = 32. Put down 32.

25/70/49
3249

Solution: The answer is 3,249.

Example 2

Find the value of 74^2.

Step 1: Find the duplexes for 7, 74, and 4 and put the three values together with slashes separating them; the slashes represent columns.

Duplex of 7: $7^2 = 49$
Duplex of 74: $2 \times 7 \times 4 = 56$
Duplex of 4: $4^2 = 16$
49/56/16

Step 2: Add the three values from right to left. You can only have one digit in every column except the last, so put down 6 and carry over the 1 to the next column.

49/56/16
6

Step 3: Add the carryover to 56: 56 + 1 = 57. Put down 7 and carry over the 5 to the next column.

49/56/16
76

Step 4: Add the carryover to 49: 49 + 5 = 54. Put down 54.

49/56/16
5476

Solution: The answer is 5,476.

Example 3

Find the value of 86^2.

Step 1: Find the duplexes for 8, 6, and 86 and put the three values together with slashes separating them; the slashes represent columns.

Duplex of 8: $8^2 = 64$
Duplex of 86: $2 \times 8 \times 6 = 96$
Duplex of 6: $6^2 = 36$
64/96/36

Step 2: Add the values from right to left. Because you can only have one digit in every column except the last, put down 6 and carry over the 3 to the next column.

64/96/36
6

Step 3: Add the carryover to 96: 96 + 3 = 99. Put down 9 and carry over the other 9 to the next column.

64/96/36
96

Step 4: Add the carryover to 64: 64 + 9 = 73. Put down 73.

64/96/36
7396

Solution: The answer is 7,396.

QUICK TIP

The duplex method for two-digit squares is an easy one to practice mentally. After you've learned it with pencil and paper, try picturing the problem in your head and the steps it takes to get the answer.

Three-Digit Squares

To find the square of any three-digit number, you need to find five duplexes: one for the first digit (a^2), one for the first and second digit ($2ab$), one for the digit as a whole ($b^2 + 2ac$), one for the second and third digit ($2ab$), and one for the third digit (a^2). Like you did with the two-digit squares, you combine the duplex values to get your answer.

Example 1

Find the value of 746^2.

Step 1: Find the duplexes for 7, 74, 746, 46, and 6 and put the five values together with slashes separating them; the slashes represent columns.

Duplex of 7: $7^2 = 49$
Duplex of 74: $2 \times 7 \times 4 = 56$
Duplex of 746: $4^2 + (2 \times 7 \times 6) = 16 + 84 = 100$
Duplex of 46: $2 \times 4 \times 6 = 48$
Duplex of 6: $6^2 = 36$
49/56/100/48/36

Step 2: Add the values from right to left. Because you can only have one digit in every column except the last, put down 6 and carry over the 3 to the next column.

49/56/100/48/36
6

Step 3: Add the carryover to 48: $48 + 3 = 51$. Put down 1 and carry over the 5 to the next column.

49/56/100/48/36
16

Step 4: Add the carryover to 100: $100 + 5 = 105$. Put down 5 and carry over the 10 to the next column.

49/56/100/48/36
516

Step 5: Add the carryover to 56: $56 + 10 = 66$. Put down 6 and carry over the 6 to the last column.

49/56/100/48/36
6516

Step 6: Add the carryover to 49: 49 + 6 = 55. Put down 55.

49/56/100/48/36

556516

Solution: The answer is 556,516.

Example 2

Find the value of 357^2.

Step 1: Find the duplexes for 3, 35, 357, 57, and 7 and put the five values together with slashes separating them; the slashes represent columns.

Duplex of 3: $3^2 = 9$
Duplex of 35: $2 \times 3 \times 5 = 30$
Duplex of 357: $5^2 + (2 \times 3 \times 7) = 25 + 42 = 67$
Duplex of 57: $2 \times 5 \times 7 = 70$
Duplex of 7: $7^2 = 49$
9/30/67/70/49

Step 2: Add the values from right to left. Because you can only have one digit in every column except the last, put down 9 and carry over the 4 to the next column.

9/30/67/70/49

9

Step 3: Add the carryover to 70: 70 + 4 = 74. Put down 4 and carry over 7 to the next column.

9/30/67/70/49

49

Step 4: Add the carryover to 67: 67 + 7 = 74. Put down 4 and carry over the 7 to the next column.

9/30/67/70/49

449

Step 5: Add the carryover to 30: 30 + 7 = 37. Put down 7 and carry over the 3 to the next column.

9/30/67/70/49

7449

Step 6: Add the carryover to 9: 9 + 3 = 12. Put down 12.

9/30/67/70/49

127449

Solution: The answer is 127,449.

Example 3

Find the value of 683^2.

Step 1: Find the duplexes for 6, 68, 683, 83, and 3 and put the five values together with slashes separating them; the slashes represent columns.

Duplex of 6: $6^2 = 36$
Duplex of 68: $2 \times 6 \times 8 = 96$
Duplex of 683: $8^2 + (2 \times 6 \times 3) = 64 + 36 = 100$
Duplex of 83: $2 \times 8 \times 3 = 48$
Duplex of 3: $3^2 = 9$
36/96/100/48/9

Step 2: Add the values from right to left. Because you only have one digit in every column except the last, you can simply put down 9.

36/96/100/48/9

9

Step 3: In the next column, put down 8 and carry over the 4 to the next column.

36/96/100/48/9

89

Step 4: Add the carryover to 100: 100 + 4 = 104. Put down 4 and carry over the 10 to the next column.

36/96/100/48/9

489

Step 5: Add the carryover to 96: 96 + 10 = 106. Put down 6 and carry over the 10 to the next column.

36/96/100/48/9

6489

Step 6: Add the carryover to 36: 36 + 10 = 46. Put down 46.

36/96/100/48/9

466489

Solution: The answer is 466,489.

Four-Digit Squares

Finding the square of a four-digit number requires to first find seven duplexes: one for the first digit (a^2); one for the first and second digit ($2ab$); one for the first, second, and third digit ($b^2 + 2ac$); one for the digit as a whole ($2ad + 2bc$); one for the second, third, and fourth digit ($b^2 + 2ac$); one for the third and fourth digit ($2ab$); and one for the fourth digit (a^2). Like you did with the two-digit and three-digit squares, you combine the duplex values to get your answer.

Remember, to get the correct answer, you must find the duplex for *all* of the combinations. So for four-digit numbers, you can't simply do the formula for four digits and expect that to be the answer; you also need the values for each digit individually.

Example 1

Find the value of $2{,}894^2$.

Step 1: Find the duplexes for 2; 28; 289; 2,894; 894; 94; and 4. Put the seven values together with slashes separating them; the slashes represent columns.

Duplex of 2: $2^2 = 4$
Duplex of 28: $2 \times 2 \times 8 = 32$
Duplex of 289: $8^2 + (2 \times 2 \times 9) = 64 + 36 = 100$
Duplex of 2894: $(2 \times 2 \times 4) + (2 \times 8 \times 9) = 16 + 144 = 160$
Duplex of 894: $9^2 + (2 \times 8 \times 4) = 81 + 64 = 145$
Duplex of 94: $2 \times 9 \times 4 = 72$
Duplex of 4: $4^2 = 16$
4/32/100/160/145/72/16

Step 2: Add the values from right to left. Because you can only have one digit in every column except the last, put down 6 and carry over the 1 to the next column.

4/32/100/160/145/72/16

6

Step 3: Add the carryover to 72: 72 + 1 = 73. Put down 3 and carry over the 7 to the next column.

4/32/100/160/145/72/16

36

Step 4: Add the carryover to 145: 145 + 7 = 152. Put down 2 and carry over the 15 to the next column.

4/32/100/160/145/72/16

236

Step 5: Add the carryover to 160: 160 + 15 = 175. Put down 5 and carry over the 17 to the next column.

4/32/100/160/145/72/16

5236

Step 6: Add the carryover to 100: 100 + 17 = 117. Put down 7 and carry over the 11 to the next column.

4/32/100/160/145/72/16

75236

Step 7: Add the carryover to 32: 32 + 11 = 43. Put down 3 and carry over the 4.

4/32/100/160/145/72/16

8375236

Step 8: Add the carryover to 4: 4 + 4 = 8. Put down 8.

4/32/100/160/145/72/16

375236

Solution: The answer is 8,375,236.

Example 2

Find the value of $1{,}234^2$.

Step 1: Find the duplexes for 1; 12; 123; 1,234; 234; 34; and 4. Put the seven values together with slashes separating them; the slashes represent columns.

Duplex of 1: $1^2 = 1$
Duplex of 12: $2 \times 1 \times 2 = 4$
Duplex of 123: $2^2 + (2 \times 1 \times 3) = 4 + 6 = 10$
Duplex of 1234: $(2 \times 1 \times 4) + (2 \times 2 \times 3) = 8 + 12 = 20$
Duplex of 234: $3^2 + (2 \times 2 \times 4) = 9 + 16 = 25$
Duplex of 34: $2 \times 3 \times 4 = 24$
Duplex of 4: $4^2 = 16$1/4/10/20/25/24/16

Step 2: Add from right to left. Because you can only have one digit in every column except the last, put down 6 carry over the 1 to the next column.

1/4/10/20/25/24/16

6

Step 3: Add the carryover to 24: 24 + 1 = 25. Put down 5 and carry over the 2.

1/4/10/20/25/24/16

56

Step 4: Add the carryover to 25: 25 + 2 = 27. Put down 7 and carry over the 2 to the next column.

1/4/10/20/25/24/16

756

Step 5: Add the carryover to 20: 20 + 2 = 22. Put down 2 and carry over the 2 to the next column.

1/4/10/20/25/24/16

2756

Step 6: Add the carryover to 10: 10 + 2 = 12. Put down 2 and carry over the 1 to the next column.

1/4/10/20/25/24/16

22756

Step 7: Add the carryover to 4: 4 + 1 = 5. Put down 5; because it's a single digit, you have no carryover.

1/4/10/20/25/24/16

522756

Step 8: Because there's no carryover to add in, simply put down 1.

1/4/10/20/25/24/16

1522756

Solution: The answer is 1,522,756.

Combined Operations: Sums of Squares

Now that you know how to find the value of individual numbers, let's take it a step further by looking at how to add squared digits together. The sum of two-digit squares is a lot like finding the duplexes for two-digit numbers, except this time you're adding the duplexes in reverse in groups. You first find the duplexes of the second digits and add them together ($a^2 + a^2$). You then find the duplexes for the numbers as a whole and add them ($2ab + 2ab$). You next find the duplexes of the first digits and add them together ($a^2 + a^2$).

The same is true for the sum of three-digit squares. You'll be using modified versions of the formulas you used earlier in reverse to find the sums—one for the third digits ($a^2 + a^2$), one for the second and third digits ($2ab + 2ab$), one for the digits as a whole ($[b^2 + 2ac] + [b^2 + 2ac]$), one for the first and second digits ($2ab$), and one for the first digits (a^2). Like with the two-digit squares, you combine the duplex values to get your answer.

You can probably guess how it goes for the sum of four-digit squares. You add the duplexes of the fourth digits ($a^2 + a^2$); the duplexes of the third and fourth digits ($2ab + 2ab$); the duplexes of the second, third, and fourth digits ($[b^2 + 2ac] + [b^2 + 2ac]$); the duplexes for the digits as a whole ($[2ad + 2bc] + [2ad + 2bc]$); the duplexes for the first, second, and third digits ($[b^2 + 2ac] + [b^2 + 2ac]$); the duplexes for the first and second digits ($2ab + 2ab$); and the duplex for the first digits ($a^2 + a^2$).

You can see how to do each type in the following examples.

<u>Example 1</u>

Solve the problem $41^2 + 33^2$.

Step 1: Find the duplexes of the second digits and add. In this case, you add the duplexes for 1 and 3.

$1^2 + 3^2 = 1 + 9 = 10$

$_10$

Step 2: Find the duplexes of the numbers as a whole and add. Here, add the duplexes of 41 and 33; then, add the carryover from the previous step.

$(2 \times 4 \times 1) + (2 \times 3 \times 3) = 8 + 18 = 26$

$26 + 1 = 27$

${}_{2}70$

Step 3: Find the duplexes for the first digits and add. In this case, you add the duplexes of 4 and 3; then, add the carryover from the previous step.

$4^2 + 3^2 = 16 + 9 = 25$

$25 + 2 = 27$

2770

Solution: The answer is 2,770.

Example 2

Solve the problem $203^2 + 112^2 + 422^2$.

Step 1: Find the duplexes of the third digits and add. In this case, you add the duplexes of 3, 2, and 2.

$3^2 + 2^2 + 2^2 = 9 + 4 + 4 = 17$

${}_{1}7$

Step 2: Find the duplexes of the second and third digits and add. Here, add the duplexes of 03, 12, and 22; then, add the carryover.

$(2 \times 0 \times 3) + (2 \times 1 \times 2) + (2 \times 2 \times 2) = 0 + 4 + 8 = 12$

$12 + 1 = 13$

${}_{1}37$

Step 3: Find the duplexes of the digits as a whole and add. In this case, you add the duplexes of 203, 112, and 422; then, add the carryover.

$(0^2 + [2 \times 2 \times 3]) + (1^2 + [2 \times 1 \times 2]) + (2^2 + [2 \times 4 \times 2]) = 12 + 5 + 20 = 37$

$37 + 1 = 38$

${}_{3}837$

Step 4: Find the duplexes of the first and second digits and add. Here, add the duplexes of 20, 11, and 42; then, add the carryover.

$(2 \times 2 \times 0) + (2 \times 1 \times 1) + (2 \times 4 \times 2) = 0 + 2 + 16 = 18$
$18 + 3 = 21$
${}_{2}1837$

Step 5: Find the duplexes of the first digits and add. In this case, add the duplexes of 2, 1, and 4; then, add the carryover.

$2^2 + 1^2 + 4^2 = 4 + 1 + 16 = 21$
$21 + 2 = 23$
231837

Solution: The answer is 231,837.

QUICK TIP

If you encounter a problem in which the numbers being added don't have the same number of digits, add zeroes. For example, for the problem $341^2 + 8^2 + 21^2$, change it to $341^2 + 008^2 + 021^2$. Adding the zeroes still gives you the same answer while also making it less complicated to use the duplex formulas.

Example 3

Solve the problem $1{,}256^2 + 4{,}765^2 + 3{,}823^2 + 5{,}995^2$.

Step 1: Find the duplexes of the fourth digits and add. Here, add the duplexes of 6, 5, 3, and 5.

$6^2 + 5^2 + 3^2 + 5^2 = 36 + 25 + 9 + 25 = 95$
${}_{9}5$

Step 2: Find the duplexes of the third and fourth digits and add. In this case, add the duplexes of 56, 65, 23, and 95; then, add the carryover.

$(2 \times 5 \times 6) + (2 \times 6 \times 5) + (2 \times 2 \times 3) + (2 \times 9 \times 5)$
$= 60 + 60 + 12 + 90 = 222$
$222 + 9 = 231$
${}_{23}15$

Step 3: Find the duplexes of the second, third, and fourth digits and add. Here, add the duplexes of 256, 765, 823, and 995; then, add the carryover.

$(5^2 + [2 \times 2 \times 6]) + (6^2 + [2 \times 7 \times 5]) + (2^2 + [2 \times 8 \times 3]) + (9^2 + [2 \times 9 \times 5]) = 49 + 106 + 52 + 171 = 378$

$378 + 23 = 401$

${}_{40}115$

Step 4: Find the duplexes of the numbers as a whole and add. In this case, add the duplexes of 1,256; 4,765; 3,823; and 5,995; then, you add the carryover.

$([2 \times 1 \times 6] + [2 \times 2 \times 5]) + ([2 \times 4 \times 5] + [2 \times 7 \times 6]) + ([2 \times 3 \times 3] + [2 \times 8 \times 2]) + ([2 \times 5 \times 5] + [2 \times 9 \times 9]) = 32 + 124 + 50 + 212 = 418$

$418 + 40 = 458$

${}_{45}8115$

Step 5: Find the duplexes of the first, second, and third digits and add. Here, add the duplexes of 125, 476, 382, and 599; then, add the carryover.

$(2^2 + [2 \times 1 \times 5]) + (7^2 + [2 \times 4 \times 6]) + (8^2 + [2 \times 3 \times 2]) + (9^2 + [2 \times 5 \times 9]) = 14 + 97 + 76 + 171 = 358$

$358 + 45 = 403$

${}_{40}38115$

Step 6: Find the duplexes of the first and second digits and add. Here, add the duplexes of 12, 47, 38, and 59; then, add the carryover.

$(2 \times 1 \times 2) + (2 \times 4 \times 7) + (2 \times 3 \times 8) + (2 \times 5 \times 9) = 4 + 56 + 48 + 90 = 198$

$198 + 40 = 238$

${}_{23}838115$

Step 7: Find the duplex of the first digits and add. In this case, add the duplexes of 1, 4, 3, and 5; then, add the carryover.

$1^2 + 4^2 + 3^2 + 5^2 = 1 + 16 + 9 + 25 = 51$

$51 + 23 = 74$

74838115

Solution: The answer is 74,838,115.

The Least You Need to Know

- For squared numbers ending in 5, use the "by one more than the one before" to get the first part of your answer.
- If you have a squared number near 50, you can find the answer by using the excess or deficit.
- The formulas a^2, $2ab$, $b^2 + 2ac$, and $2ad + 2bc$ give you the duplex of one digit, two digits, three digits, and four digits respectively.
- When adding squared numbers, you simply group by duplex and add.

CHAPTER

11

Cubed Numbers

In This Chapter

- Learning how to find the cube of any two-digit number
- Finding the cubes of numbers near a base
- Applying the ratio $\frac{b}{a}$ to find the cube of a number

Now that you know how to find squares, let's move on to cubes. Like with squared numbers, it's not difficult to put aside the calculator to find the answers once you know some quick and interesting shortcuts.

In this chapter, I talk about finding the value of two-digit cubed numbers and cubed numbers near a base.

Finding the Cubes of Two-Digit Numbers

Do you know the formula for the expansion of $(a + b)^3$? This formula is important, as it will help you find the cubes of various numbers. Let's see the formula:

$$
\begin{array}{rl}
(a+b)^3 = & a^3 + a^2b + ab^2 + b^3 \\
 & \quad 2a^2b + 2ab^2 \\
\hline
 & a^3 + 3a^2b + 3ab^2 + b^3
\end{array}
$$

In this formula, consider a to be in the tens place and b the ones place. If you take a closer look at the first line, you can spot a ratio between a^3 and a^2b; if you divide a^2b by a^3, you get the ratio $\frac{b}{a}$. Similarly, you get $\frac{b}{a}$ when you divide ab^2 by a^2b and ab^2 by b^3. The ratio $\frac{b}{a}$ is very helpful in calculating the complete value of the cube. We multiply each term in the first line by this ratio.

In the second line, a^2b has been doubled to $2a^2b$ and ab^2 has been doubled to $2ab^2$. Added together, these two lines give you the expression for $(a + b)^3$, which is $a^3 + 3a^2b + 3ab^2 + b^3$. That means you can find the numbers for the two lines and combine them to find a two-digit number to the third power.

Take a look at the following examples to see how you can apply the formula to find the cube of a two-digit number.

Example 1

Find the value of 13^3.

Step 1: Here, a is 1 and b is 3. Cube a: $1^3 = 1$. The 1 is the first number in the line.

$13^3 = 1$

Step 2: Find the rest of the numbers for the first line as set up in the expansion of $(a + b)^3$. Because $\frac{b}{a}$ is $\frac{3}{1}$, multiply each subsequent digit by 3 until you get to the b^3 value: $1 \times 3 = 3$, $3 \times 3 = 9$, and $9 \times 3 = 27$. The last number, 27, is equal to b^3, so the first line is complete.

$13^3 = 1 \quad 3 \quad 9 \quad 27$

Step 3: To get the second line, multiply the second and third terms by 2, as you saw in the expansion: $3 \times 2 = 6$ and $9 \times 2 = 18$.

$$
\begin{array}{llll}
13^3 = 1 & 3 & 9 & 27 \\
 & 6 & 18 &
\end{array}
$$

Step 4: Start adding from right to left, making sure every column before the last only has one digit in it. In the first column, bring down the 7 and carry over the 2 to the next step.

$$\begin{array}{rl} 13^3 = & 1\ \ 3\ \ 9\ \ 27 \\ & +\ 6\ \ 18 \\ \hline & \qquad\quad 7 \end{array}$$

Step 5: In the next column, add 9, 18, and the carryover: 9 + 18 + 2 = 29. Put down 9 and carry over the 2 to the next step.

$$\begin{array}{rl} 13^3 = & 1\ \ 3\ \ 9\ \ 27 \\ & +\ 6\ \ 18 \\ \hline & \qquad\ 97 \end{array}$$

Step 6: In the next column, add 3, 6, and the carryover: 3 + 6 + 2 = 11. Put down 1 and carry over the 1 to the next step.

$$\begin{array}{rl} 13^3 = & 1\ \ 3\ \ 9\ \ 27 \\ & +\ 6\ \ 18 \\ \hline & \quad\ 197 \end{array}$$

Step 7: In the last column, add 1 and the carryover: 1 + 1 = 2. Put down 2.

$$\begin{array}{rl} 13^3 = & 1\ \ 3\ \ 9\ \ 27 \\ & +\ 6\ \ 18 \\ \hline & \ \ 2197 \end{array}$$

Solution: The answer is 2,197.

QUICK TIP

Remember, when getting cubes, each column before the last should have only one digit. That will make your addition very easy.

<u>**Example 2**</u>

Find the value of 32^3.

Step 1: Here, *a* is 3 and *b* is 2. Cube *a*: $3^3 = 27$. The 27 is the first number in the line.

$32^3 = 27$

Step 2: Find the rest of the numbers for the first line as set up in the expansion of $(a + b)^3$. Because $\frac{b}{a}$ is $\frac{2}{3}$, multiply each subsequent digit by 2 and divide by 3 until you get to the b^3 value:

$(27 \times 2) \div 3 = 18$, $(18 \times 2) \div 3 = 12$, and $(12 \times 2) \div 3 = 8$. The last number, 8, is equal to b^3, so the first line is complete.

$$32^3 = 27 \quad 18 \quad 12 \quad 8$$

Step 3: To get the second line, multiply the second and third terms by 2, as you saw in the expansion: $18 \times 2 = 36$ and $12 \times 2 = 24$.

$$\begin{array}{lllll} 32^3 = 27 & 18 & 12 & 8 \\ & 36 & 24 & \end{array}$$

Step 4: Start adding from right to left, making sure every column before the last only has one digit in it. In the first column, bring down 8.

$$\begin{array}{r} 32^3 = 27\ 18\ 12\ 8 \\ +\ 36\ 24\ \ \\ \hline 8 \end{array}$$

Step 5: In the next column, add 12 and 24: $12 + 24 = 36$. Put down 6 and carry over the 3 to the next step.

$$\begin{array}{r} 32^3 = 27\ 18\ 12\ 8 \\ +\ 36\ 24\ \ \\ \hline 68 \end{array}$$

Step 6: In the next column, add 18, 36, and the carryover: $18 + 36 + 3 = 57$. Put down 7 and carry over the 5 to the next step.

$$\begin{array}{r} 32^3 = 27\ 18\ 12\ 8 \\ +\ 36\ 24\ \ \\ \hline 768 \end{array}$$

Step 7: In the last column, add 27 and the carryover: $27 + 5 = 32$. Put down 32.

$$\begin{array}{r} 32^3 = 27\ 18\ 12\ 8 \\ +\ 36\ 24\ \ \\ \hline 32768 \end{array}$$

Solution: The answer is 32,768.

<u>Example 3</u>

Find the value of 38^3.

Step 1: Here, a is 3 and b is 8. Cube a: $3^3 = 27$. The 27 is the first number in the line.

$38^3 = 27$

Step 2: Find the rest of the numbers for the first line as set up in the expansion of $(a + b)^3$. Because $\frac{b}{a}$ is $\frac{8}{3}$, multiply each subsequent digit by 8 and divide by 3 until you get to the b^3 value: $(27 \times 8) \div 3 = 72$, $(78 \times 8) \div 3 = 192$, and $(192 \times 8) \div 3 = 512$. The last number, 512, is equal to b^3, so the first line is complete.

$38^3 = 27 \quad 72 \quad 192 \quad 512$

Step 3: To get the second line, multiply the second and third terms by 2, as you saw in the expansion: $72 \times 2 = 144$ and $192 \times 2 = 384$.

$$\begin{array}{llll} 38^3 = 27 & 72 & 192 & 512 \\ & 144 & 384 & \end{array}$$

Step 4: Start adding from right to left, making sure every column before the last only has one digit in it. In the first column, bring down 2 and carry over the 51 to the next step.

$$\begin{array}{rrrr} 38^3 = 27 & 72 & 192 & 512 \\ & + 144 & 384 & \\ \hline & & & 2 \end{array}$$

Step 5: In the next column, add 192, 384, and the carryover: $192 + 384 + 51 = 627$. Put down 7 and carry over the 62 to the next step.

$$\begin{array}{rrrr} 38^3 = 27 & 72 & 192 & 512 \\ & + 144 & 384 & \\ \hline & & & 72 \end{array}$$

Step 6: In the next column, add 72, 144, and the carryover: $72 + 144 + 62 = 278$. Put down 8 and carry over the 27 to the next step.

$$\begin{array}{rrrr} 38^3 = 27 & 72 & 192 & 512 \\ & + 144 & 384 & \\ \hline & & & 872 \end{array}$$

Step 7: In the last column, add 27 and the carryover: 27 + 27 = 54. Put down 54.

$$
\begin{array}{l}
38^3 = 27\ \ 72\ \ 192\ \ 512 \\
\quad\quad\quad + 144\ \ 384 \\
\hline
\quad\quad\quad\quad 54872
\end{array}
$$

Solution: The answer is 54,872.

Calculating Cubes Near a Base

If the number being cubed is near a base of 100; 1,000; 10,000; and so on, there's a process you can use to more easily find the answer. First, you need to know the cubes of numbers 1 through 9, as you'll be using them in your calculations.

Number	Cubes
1	1
2	8
3	27
4	64
5	125
6	216
7	343
8	512
9	729

Now that you've looked over the cube values, the following sections detail how to find the cube depending on whether your number is above or below the base.

QUICK TIP

At some point, take the time to memorize these values. The cube of a single-digit number is part of the process, so having them at the ready will make that part of solving the problem a breeze.

When the Number Is Above the Base

To find the cube of a number above the base, you add the number to two times its excess of the base to get the first part of your answer. You then square the excess and multiply it by 3 to get the second part. To get the third and final part, you find the cube of the excess.

Example 1

Find the value of 105^3.

Step 1: This number is closest to a base of 100; the excess is 5, as 105 is 5 more than the base. To get the first part of the answer, add 105 to two times the excess: $105 + (5 \times 2) = 115$.

$105^3 = 115$

Step 2: To get the second part of the answer, square the excess: $5^2 = 25$. Multiply that number by 3: $25 \times 3 = 75$.

$105^3 = 115/75$

Step 3: To find the third and final part of the answer, find the cube of the excess (see the earlier table if you need a refresher): $5^3 = 125$.

$105^3 = 115/75/125$

Step 4: Combine the three parts. The second and third part of the answer should only contain the number of digits equal to the number of zeroes in the base. In this case, because the base is 100, there should only be two digits in those parts. That means, for the third part, you need to carry over the 1 and add it to the second part: $75 + 1 = 76$.

$105^3 = 115/75/125$

$105^3 = 115/76/25 = 1157625$

Solution: The answer is 1,157,625.

Example 2

Find the value of $1{,}009^3$.

Step 1: This number is closest to a base of 1,000; the excess is 9, as 1,009 is 9 more than the base. To get the first part of the answer, add 1,009 to two times the excess: $1{,}009 + (9 \times 2) = 1{,}027$.

$1009^3 = 1027$

Step 2: To get the second part of the answer, square the excess: $9^2 = 81$. Multiply that number by 3: $81 \times 3 = 243$.

$1009^3 = 1027/243$

Step 3: To find the third and final part of the answer, find the cube of the excess (see the earlier table if you need a refresher): $9^3 = 729$.

$1009^3 = 1027/243/729$

Step 4: Combine the three parts. The second and third part of the answer should only contain the number of digits equal to the number of zeroes in the base. In this case, because the base is 1,000, there should only be three digits in those parts. Both parts already have three digits, so you don't need to do any carryovers.

$1009^3 = 1027/243/729 = 1027243729$

Solution: The answer is 1,027,243,729.

<u>Example 3</u>

Find the value of $10{,}006^3$.

Step 1: This number is closest to a base of 10,000; the excess is 6, as 10,009 is 6 more than the base. To get the first part of the answer, add 10,006 to two times the excess: $10{,}006 + (6 \times 2)$ 10,018.

$10006^3 = 10018$

Step 2: To get the second part of the answer, square the excess: $6^2 = 36$. Multiply that number by 3: $36 \times 3 = 108$.

$10006^3 = 10018/108$

Step 3: To find the third and final part of the answer, find the cube of the excess (see the earlier table if you need a refresher): $6^3 = 216$.

$10006^3 = 10018/108/216$

Step 4: Combine the three parts. The second and third part of the answer should only contain the number of digits equal to the number of zeroes in the base. Because the base is 10,000, there should be four digits in those parts. In this case, you need to add zeroes before the 108 and 216 to make them four digits.

$10006^3 = 10018/108/216$
$10006^3 = 10018/0108/0216 = 1001801080216$

Solution: The answer is 1,001,801,080,216.

When the Number Is Below the Base

For a number below the base, you have to change up the process a bit. The deficit is treated as a negative number, so instead of adding in the first part, you subtract. The second part is exactly the same process—you square the difference and multiply it by 3. The process for getting the third part is the same as well—you cube the difference—but what you do with it when combining the parts is altered. The third part of the answer will be a negative number, so in order to get rid of the negative, you carry over the base from the second part and subtract the third part from that.

Example 1

Find the value of 96^3.

Step 1: This number is closest to a base of 100; the deficit is 4, as 96 is 4 less than the base. To get the first part of the answer, subtract 96 by two times the deficit: $96 - (4 \times 2) = 88$.

$96^3 = 88$

Step 2: To get the second part of the answer, square the deficit: $-4^2 = 16$. Multiply that number by 3: $16 \times 3 = 48$.

$96^3 = 88/48$

Step 3: To find the third and final part of the answer, find the cube of the excess (see the earlier table if you need a refresher; this version has a negative): $-4^3 = -64$.

$96^3 = 88/48/-64$

Step 4: Combine the three parts. You can't have a negative number in the answer, so for the third part, carry over base 100 from the second part. That simply means subtracting 1 from 48, as 8 is in the hundreds place: $48 - 1 = 47$. You then subtract the third part from 100: $100 - 64 = 36$.

$96^3 = 88/48/-64$

$96^3 = 88/47/36 = 884736$

Solution: The answer is 884,736.

Example 2

Find the value of 992^3.

Step 1: This number is closest to a base of 1,000; the deficit is 8, as 992 is 8 less than the base. To get the first part of the answer, subtract 992 by two times the deficit: $992 - (8 \times 2) = 976$.

$992^3 = 976$

Step 2: To get the second part of the answer, square the deficit: $-8^2 = 64$. Multiply that number by 3: $64 \times 3 = 192$.

$992^3 = 976/192$

Step 3: To find the third and final part of the answer, find the cube of the excess (see the earlier table if you need a refresher; this version has a negative): $-8^3= -512$.

$992^3 = 976/192/-512$

Step 4: Combine the three parts. You can't have a negative number in the answer, so for the third part, carry over base 1,000 from the second part. That simply means subtracting 1 from 192, as 2 is in the thousands place: $192 - 1 = 191$. You then subtract the third part from 1,000: $1,000 - 512 = 488$.

$992^3 = 976/192/-512$
$992^3 = 976/191/488 = 976191488$

Solution: The answer is 976,191,488.

Example 3

Find the value of $9,993^3$.

Step 1: This number is closest to a base of 10,000; the deficit is 7, as 9,993 is 7 less than the base. To get the first part of the answer, subtract 9,993 by two times the deficit: $9,993 - 14 = 9,979$.

$9993^3 = 9979$

Step 2: To get the second part of the answer, square the deficit: $-7^2 = 49$. Multiply that number by 3: $49 \times 3 = 147$. Write this as 0147 because of the placement rule.

$9993^3 = 9979/0147$

QUICK TIP

The placement rule states that the number of digits in the column or section should match the number of zeroes in the base. For base 10,000, there are four zeroes; therefore, 147 is written as 0147 to make it a four-digit number.

Step 3: To find the third and final part of the answer, find the cube of the excess (see the earlier table if you need a refresher; this version has a negative): $-7^3 = -343$. Write this as -0343 because of the placement rule.

$9993^3 = 9979/0147/\text{-}0343$

Step 4: Combine the three parts. You can't have a negative number in the answer, so for the third part, carry over base 10,000 from the second part. That simply means subtracting 1 from 0147, as 7 is in the ten thousands place: 0147 – 1 = 0146. You then subtract the third part from 10,000: 10,000 – 0343 = 9,657.

$9993^3 = 9979/0147/\text{-}0343$

$9993^3 = 9979/0146/9657 = 997901469657$

Solution: The answer is 997,901,469,657.

The Least You Need to Know

- To find out how much a number is cubed, use $a^3 + a^2b + a^2b^2 + ab^2 + b^3$ and $2a^2b + 2ab^2$ to find the values, and add right to left.
- Use $\frac{b}{a}$ to find the numbers for the first line of the formula.
- If the number being cubed is above or below a base, you can use the excess or deficit to help you find the value.

CHAPTER

12

Numbers to the Fourth and Fifth Powers

In This Chapter

- How to calculate a number to the fourth power
- How to calculate a number to the fifth power
- The significance of the ratio $\frac{b}{a}$

The ways to calculate a number to the fourth and fifth power aren't so different from what you learned from squares and cubes in Chapters 10 and 11. I think you'll be quite surprised at how quickly you pick up the formulas for fourth and fifth powers. You may even find yourself not even needing to write things down!

In this chapter, I give you the formulas for calculating the values of numbers to the fourth and fifth powers and show you how to apply them.

Calculating a Number to the Fourth Power

Like you did for cubes, to find out what a number is to the fourth power, you first plug in the digits to $(a + b)$. Again, a is the tens place of the number and b is the ones place of the number. Let's see what the expanded formula for $(a + b)$ is when raised to the fourth power:

$$(a+b)^4 = a^4 + a^3b + a^2b^2 + ab^3 + b^4$$
$$3a^3b + 5a^2b^2 + 3ab^3$$
$$\overline{a^4 + 4a^3b + 6a^2b^2 + 4ab^3 + b^4}$$

Like the cube formula, the formula for the fourth power contains the ratio $\frac{b}{a}$. You get that by dividing a^3b by a^4, a^2b^2 by a^3b, and ab^3 by b^4. If you recall from Chapter 11, you multiply each term in the first line by this ratio to help you find the value of a number to the fourth power.

Now consider the second line. Here, a^3b and ab^3 have been tripled to $3a^3b$ and $3ab^3$, and a^2b^2 has been multiplied five times to make it $5a^2b^2$. Added together, these two lines give you the expression for $(a + b)^4$, which is $a^4 + 4a^3b + 6a^2b^2$ $4ab^3 + b^4$. That means you can find the numbers for the two lines and combine them to find a two-digit number to the fourth power.

The following are some examples that show you how to plug in to the expanded formula to get the value of a number raised to the fourth power.

<u>Example 1</u>

Find the value of 12^4.

Step 1: Here, a is 1 and b is 2, so $\frac{b}{a}$ is $\frac{2}{1}$, or 2. Raise a to the fourth power: $1^4 = 1$. This 1 is the first number in the line.

$12^4 = 1$

Step 2: Find the rest of the numbers for the first line as set up in the expansion of $(a + b)^4$. Multiply 1 and $\frac{b}{a}$: $1 \times 2 = 2$. Place the 2 to the right of the 1. Because $\frac{b}{a}$ is $\frac{2}{1}$, multiply each subsequent digit by 2 until you get to the b^4 value: $2 \times 2 = 4$, $4 \times 2 = 8$, and $8 \times 2 = 16$. Because 16 is equal to b^4, your first line is complete.

$12^4 = 1 \quad 2 \quad 4 \quad 8 \quad 16$

Step 3: To get the second line, multiply the second term by 3, the third term by 5, and the fourth term by 3, as you saw in the expansion: 2 × 3 = 6, 4 × 5 = 20, and 8 × 3 = 24.

$$
\begin{array}{lllll}
12^4 = 1 & 2 & 4 & 8 & 16 \\
 & 6 & 20 & 24 &
\end{array}
$$

Step 4: Start adding from right to left, making sure every column before the last only has one digit in it. In the first column, put down 6 and carry over the 1 to the next step.

$$
\begin{array}{r}
12^4 = 1\ 2\ 4\ 8\ 16 \\
6\ 20\ 24 \\
\hline
6
\end{array}
$$

Step 5: In the next column, add 8, 24, and the carryover: 8 + 24 + 1 = 33. Put down 3 and carry over the 3 to the next step.

$$
\begin{array}{r}
12^4 = 1\ 2\ 4\ 8\ 16 \\
6\ 20\ 24 \\
\hline
3\ 6
\end{array}
$$

Step 6: In the next column, add 4, 20, and the carryover: 4 + 20 + 3 = 27. Put down 7 and carry over the 2 to the next step.

$$
\begin{array}{r}
12^4 = 1\ 2\ 4\ 8\ 16 \\
6\ 20\ 24 \\
\hline
7\ 3\ 6
\end{array}
$$

Step 7: In the next column, add the 2, 6, and carryover: 2 + 6 + 2 = 10. Put down 0 and carry over the 1 to the next step.

$$
\begin{array}{r}
12^4 = 1\ 2\ 4\ 8\ 16 \\
6\ 20\ 24 \\
\hline
0\ 7\ 3\ 6
\end{array}
$$

Step 8: In the last column, add 1 and the carryover: 1 + 1 = 2. Put down 2.

$$
\begin{array}{r}
12^4 = 1\ 2\ 4\ 8\ 16 \\
6\ 20\ 24 \\
\hline
2\ 0\ 7\ 3\ 6
\end{array}
$$

Solution: The answer is 20,736.

QUICK TIP

If you aren't sure what would be considered the b^4 value and therefore don't know when to stop multiplying, think of it this way: you only need five values for the first line. So once you get to the fifth number, you can move on to finding the numbers for the second line.

<u>Example 2</u>

Find the value of 13^4.

Step 1: Here, *a* is 1 and *b* is 3. Take *a*, which is 1, and raise it to the fourth power: $1^4 = 1$. This 1 is the first number in the line.

$13^4 = 1$

Step 2: Find the rest of the numbers for the first line as it is in the expansion of $(a + b)^4$. Because $\frac{b}{a}$ is $\frac{3}{1}$, you can get the first line of the problem by multiplying each subsequent number by 3 until you get the b^4 value: 1 × 3 = 3, 3 × 3 = 9, 9 × 3 = 27, and 27 × 3 = 81. The last number, 81, is equal to b^4, so the first line is complete.

$$13^4 = 1 \quad 3 \quad 9 \quad 27 \quad 81$$

Step 3: To get the second line, multiply the second term by 3, the third term by 5, and the fourth term by 3, as you saw in the expansion: 3 × 3 = 9, 9 × 5 = 45, and 27 × 3 = 81.

$$
\begin{array}{lllll}
13^4 = 1 & 3 & 9 & 27 & 81 \\
 & 9 & 45 & 81 &
\end{array}
$$

Step 4: Add from right to left, making sure every column before the last only has one digit in it. In the first column, for 81, put down 1 and carry over the 8 to the next step.

				8	
13^4	= 1	3	9	27	81
		9	45	81	
					1

Step 5: In the next column, add 27, 81, and the carryover: 27 + 81 + 8 = 116. Put down 6 and carry over the 11 to the next step.

			11	8	
13^4	= 1	3	9	27	81
		9	45	81	
				6	1

Step 6: In the next column, add 9, 45, and the carryover: 9 + 45 + 11 = 65. Put down 5 and carry over the 6 to the next step.

		6	11	8	
13^4	= 1	3	9	27	81
		9	45	81	
			5	6	1

Step 7: In the next column, add 3, 9, and the carryover: 3 + 9 + 6 = 18. Put down 8 and carry over the 1 to the next step.

	1	6	11	8	
13^4	= 1	3	9	27	81
		9	45	81	
		8	5	6	1

Step 8: In the last column, add 1 and the carryover: 1 + 1 = 2. Put down 2.

	1	6	11	8	
13^4	= 1	3	9	27	81
		9	45	81	
	2	8	5	6	1

Solution: The answer is 28,561.

Example 3

Find the value of 32^4.

Step 1: Here, *a* is 3 and *b* is 2. Take *a*, which is 3, and raise it to the fourth power: $3^4 = 81$. The 81 is the first number in the line.

$32^4 = 81$

Step 2: Find the rest of the numbers for the first line as it is in the expansion of $(a + b)^4$ until you get to the b^4 value. Because $\frac{b}{a}$ is $\frac{2}{3}$, you can get our first line of the problem by multiplying each subsequent number by 2 and dividing by 3: $(81 \times 2) \div 3 = 54$, $(54 \times 2) \div 3 = 36$, $(36 \times 2) \div 3 = 24$, and $(24 \times 2) \div 3 = 16$. The last number, 16, is equal to b^4, so the first line is complete.

$32^4 = 81 \quad 54 \quad 36 \quad 24 \quad 16$

Step 3: To get the second line, multiply the second term by 3, the third term by 5, and the fourth term by 3, as you saw in the expansion: $54 \times 3 = 162$, $36 \times 5 = 180$, and $24 \times 3 = 72$.

$32^4 = 81$	54	36	24	16
	162	180	72	

Step 4: Add right to left, making sure every column before the last only has one digit in it. In the first column, for 16, put down 6 and carry over the 1 to the next step.

32^4 =	81	54	36	$\overset{1}{24}$	16
		162	180	72	
					6

Step 5: In the next column, add 24, 72, and the carryover: $24 + 72 + 1 = 97$. Put down 7 and carry over the 9 to the next step.

32^4 =	81	54	$\overset{9}{36}$	$\overset{1}{24}$	16
		162	180	72	
				7	6

Step 6: In the next column, add 36, 180, and the carryover: 36 + 180 + 9 = 225. Put down 5 and carry over the 22 to the next step.

		22	9	1	
$32^4 =$	81	54	36	24	16
		162	180	72	
			5	7	6

Step 7: In the next column, add 54, 162, and the carryover: 54 + 162 + 22 = 238. Put down 8 and carry over the 23 to the next step.

	23	22	9	1	
$32^4 =$	81	54	36	24	16
		162	180	72	
		8	5	7	6

Step 8: In the last column, add 81 and the carryover: 81 + 23 = 104. Put down 104.

	23	22	9	1	
$32^4 =$	81	54	36	24	16
		162	180	72	
	104	8	5	7	6

Solution: The answer is 1,048,576.

Calculating a Number to the Fifth Power

The method for calculating what a number is to the fifth power is very much like what you did for the fourth power. First, let's see what the expanded formula is for $(a + b)^5$:

$$(a + b)^5 = a^5 + 5a^4b + 10a^3b^2 + 10a^2b^3 + 5ab^4 + b^5$$

Broken down further, here's what the first line looks like:

$$a^5 + a^4b + a^3b^2 + a^2b^3 + ab^4 + b^5$$

Like the formula for the fourth power, the ratio of $\frac{b}{a}$ holds true here as well.

The second line of the formula is this:

$4a^4b + 9a^3b^2 + 9a^2b^3 + 4ab4$

So $(a + b)^5$ should look like the following:

$$
\begin{array}{ll}
(a+b)^5 = & a^5 + a^4b + a^3b^2 + a^2b^3 + ab^4 + b^5 \\
 & \quad 4a^4b + 9a^3b^2 + 9a^2b^3 + 4ab^4 \\
\hline
 & a^5 + 5a^4b + 10a^3b^2 + 10a^2b^3 + 5ab^4 + b^5
\end{array}
$$

And that's it! Try out this formula by working through the following examples.

<u>Example 1</u>

Find the value of 12^5.

Step 1: Here, a is 1 and b is 2. Take the a value and raise it to the fifth power: $1^5 = 1$. The 1 is the first number in the line.

$12^5 = 1$

Step 2: Find the rest of the numbers for the first line as it is in the expansion of $(a + b)^5$ until you get to the b^5 value. Because $\frac{b}{a}$ is $\frac{2}{1}$, multiply each subsequent number by 2: $1 \times 2 = 2$, $2 \times 2 = 4$, $4 \times 2 = 8$, $8 \times 2 = 16$, and $16 \times 2 = 32$. The last number, 32, is equal to b^5, so the first line is complete.

$12^5 = 1 \quad 2 \quad 4 \quad 8 \quad 16 \quad 32$

Step 3: To get the second line, multiply the second term by 4, the third term by 9, and the fourth term by 9, and the fifth term by 4, as you saw in the expansion: $2 \times 4 = 8$, $4 \times 9 = 36$, $8 \times 9 = 72$, and $16 \times 4 = 64$.

$$
\begin{array}{llllll}
12^5 = 1 & 2 & 4 & 8 & 16 & 32 \\
 & 8 & 36 & 72 & 64 &
\end{array}
$$

Step 4: Add right to left, making sure every column before the last only has one digit in it. In the first column, put down 2 and carry over the 3 to the next step.

```
                          3
12^5 = 1   2    4    8   16   32
           8   36   72   64
--------------------------------
                               2
```

Step 5: In the next column, add 16, 64, and the carryover: 16 + 64 + 3 = 83. Put down 3 and carry over the 8 to the next step.

```
                     8    3
12^5 = 1   2    4    8   16   32
           8   36   72   64
--------------------------------
                          3    2
```

Step 6: In the next column, add 8, 72, and the carryover: 8 + 72 + 8 = 88. Put down 8 and carry over the other 8 to the next step.

```
                8    8    3
12^5 = 1   2    4    8   16   32
           8   36   72   64
--------------------------------
                     8    3    2
```

Step 7: In the next column, add 4, 36, and the carryover: 4 + 36 + 8 = 48. Put down 8 and carry over the 4 to the next step.

```
           4    8    8    3
12^5 = 1   2    4    8   16   32
           8   36   72   64
--------------------------------
                8    8    3    2
```

Step 8: In the next column, add 2, 8, and the carryover: 2 + 8 + 4 = 14. Put down 4 and carry over the 1 to the next step.

```
       1   4    8    8    3
12^5 = 1   2    4    8   16   32
           8   36   72   64
--------------------------------
           4    8    8    3    2
```

Step 9: In the last column, add 1 and the carryover: 1 + 1 = 2. Put down 2.

	1	4	8	8	3	
$12^5 =$	1	2	4	8	16	32
		8	36	72	64	
	2	4	8	8	3	2

Solution: The answer is 248,832.

SPEED BUMP

Don't forget, if the *a* digit isn't 1, you can't simply multiply by *b* and have your answer; because you're using the ratio $\frac{b}{a}$, you have to multiply by *b* *and* divide by *a*.

The Least You Need to Know

- To find out how much a number is to the fourth power, use $a^4 + a^3b + a^2b^2 + ab^3 + b^4$ and $3a^3b + 5a^2b^2 + 3ab^3$ to find the values, and add right to left.
- To do fifth power calculations, use $a^5 + a^4b + a^3b^2 + a^2b^3 + ab^4 + b^5$ and $4a^4b + 9a^3b^2 + 9a^2b^3 + 4ab^4$ to find the values, and add right to left.
- Use $\frac{b}{a}$ to find the numbers for the first line of the formula.

CHAPTER

13

Square and Cube Roots

In This Chapter

- Finding the square root of perfect squares
- Getting the square root of any given number
- Discovering the value of perfect cubes

Seeing the square root symbol on a page can inspire nervousness. In some cases, it can be a guessing game, just trying to multiply a number by itself to see if it's even close. And then there's the old standby—the calculator. But finding square or even cube roots doesn't have to be so frustrating.

In this chapter, you learn some ways to painlessly find the square and cube roots of numbers.

Square Roots of Perfect Squares

To begin the discussion of roots, let me show you how to find square roots for perfect squares. A perfect square is a number whose square root is an integer, or whole number. So for the examples in this section, all of your answers should be whole numbers. Not too hard, right?

Take a look at the following table to see what constitutes a perfect square.

Number	Square	Last Digit	Digit Sum
1	1	1	1
2	4	4	4
3	9	9	9
4	16	6	7
5	25	5	7
6	36	6	9
7	49	9	4
8	64	4	1
9	81	1	9

You should notice two things about the squares for these numbers:

1. They have a digit sum (see Chapter 5) of 1, 4, 7, or 9.
2. They end in 1, 4, 5, 6, or 9.

So right away, you have two indicators that can help you determine whether a number is a perfect square.

To find the value of a perfect square, you split the number into groups of two to see how many digits are in the answer. You then look at the early digits to see which square value they're between, giving you the first part of the answer. The table can help you with the last couple steps, which involve knowing the number that has the same last digit when squared and finding the digit sums of the number and the possible answers.

Example 1

Find $\sqrt{3249}$.

Step 1: Split the number into groups of two to determine how many digits are in the answer. Because you get two groups, 32 and 49, the number of digits in the answer is going to be two.

Step 2: Look at the first two digits of the number together and figure out which square values it's closest to. In this case, 32 is more than 25 (5^2) and less than 36 (6^2). This group is the tens place of your answer, so the first part of the answer is between 50 and 60.

Step 3: Look at the last digit of 3,249. If you recall from the table, 3 and 7 have a last digit of 9 when squared. Therefore, the square is either 53 or 57.

Step 4: Compare the digit sums of 53 and 57 squared to the digit sum of 3,249 to see which one matches.

$3249: 3 + 2 + 4 + 9 = 18; 1 + 8 = 9$
$53^2: (5 + 3)^2 = 8^2 = 64; 6 + 4 = 10; 1 + 0 = 1$
$57^2: (5 + 7)^2 = 12^2 = 144; 1 + 4 + 4 = 9$

Solution: The answer is 57.

<u>Example 2</u>

Find $\sqrt{9604}$.

Step 1: Split the number into groups of two to determine how many digits are in the answer. You get two groups, 96 and 04, so the answer's going to be two digits.

Step 2: Look at the first two digits of the number together and figure out which square values it's closest to. In this case, 96 is more than 81 (9^2) and less than 100 (10^2). This group is the tens place of your answer, so the first part of the answer is between 90 and 100.

Step 3: Look at the last digit of 9,604. If you recall from the table, 2 and 8 have a last digit of 4 when squared. Therefore, the square is either 92 or 98.

Step 4: Compare the digit sums of 92 and 98 squared to the digit sum of 9,604 to see which one matches.

$9604: 9 + 6 + 0 + 4 = 19; 1 + 9 = 10; 1 + 0 = 1$
$92^2: (9 + 2)^2 = 11^2 = 121; 1 + 2 + 1 = 5$
$98^2: (9 + 8)^2 = 17^2 = 289; 2 + 8 + 9 = 19; 1 + 9 = 10;$
$1 + 0 = 1$

Solution: The answer is 98.

QUICK TIP

What do you do when the digit sum for the potential answers is the same? If this happens, simply find the square of a number between them and see how it compares. For example, the square root of 2,401 is either 41 and 49, both of which have a digit sum of 7. To get the right answer, find the value of 45^2. Its value is 2,025, which is lower than the original number. That means the square root of 2,401 is 49.

<u>Example 3</u>

Find $\sqrt{24964}$.

Step 1: Split the number into groups of two to determine how many digits are in the answer. For this number, you have five digits, so get two groups of two and one group of one, meaning there are going to be three digits in the answer.

Step 2: Look at the first three digits of the number together and figure out which square values it's closest to. In this case, 249 is more than 225 (15^2) and less than 256 (16^2), so the first two digits of the answer must be 15.

Step 3: Look at the last digit of 24,964. If you recall from the table, 2 and 8 have a last digit of 4 when squared. Therefore, the square is either 152 or 158.

Step 4: Compare the digit sums of 152 and 158 squared to the digit sum of 24,964 to see which one matches.

$24964: 2 + 4 + 9 + 6 + 4 = 25; 2 + 5 = 7$
$152^2: (1 + 5 + 2)^2 = 8^2 = 64; 6 + 4 = 10; 1 + 0 = 1$
$158^2: (1 + 5 + 8)^2 = 14^2 = 196; 1 + 9 + 6 = 16; 1 + 6 = 7$

Solution: The answer is 158.

Using Duplexes to Find Square Roots

You can find the square root of any number by cycling through two steps:

1. Divide by 2 times the first digit of your answer.
2. Subtract by the duplexes (see Chapter 10 if you need to jog your memory).

Let me show you how it's done with some examples.

<u>**Example 1**</u>

Find $\sqrt{529}$.

Step 1: Divide the number into groups of two from right to left. Because 529 is a three-digit number, you have one group of two (29) and one group of one (5). There are two groups, so the answer is going to have two digits before the decimal.

$$\begin{array}{c|l} & \overline{5}\,\overline{29} \\ \hline & \end{array}$$

Step 2: For the first group, 5, find the perfect square just less than it. The closest perfect square that's less than 5 is 4. The square root of 4 is the first digit of your answer, so put down 2. To get the divisor, double 2: 2 × 2 = 4. Put the 4 on the left.

$$\begin{array}{c|l} 4 & \overline{5}\,\overline{29} \\ \hline & 2 \end{array}$$

Step 3: Subtract 5 minus the square of 2. So we get 5 – 4 = 1. This 1 is the remainder which is then prefixed to 2, making it 12.

$$\begin{array}{c|l} 4 & \overline{5}_{1}\overline{29} \\ \hline & 2 \end{array}$$

Step 4: Divide 12 by 4: 12 ÷ 4 = 3, remainder 0. Put down 3 and prefix the remainder to the 9 so it becomes 09.

$$\begin{array}{c|l} 4 & \overline{5}_{1}\overline{2_{0}9} \\ \hline & 23 \end{array}$$

Step 5: Subtract 09 by the duplex of 3. In this case, the duplex of 3 is 3^2, or 9: 09 – 9 = 0. Dividing 0 by 4 gives you 0, so you're finished. Because the answer only has two digits, the 0 goes after the decimal.

$$\begin{array}{r|l} 4 & \overline{5}_1\overline{2_0 9} \\ \hline & 23.0 \end{array}$$

Solution: The answer is 23.0.

<u>Example 2</u>

Find $\sqrt{4624}$.

Step 1: Divide the number into groups of two from right to left. Because there are two groups, 46 and 24, the answer is going to contain two digits before the decimal.

$$\begin{array}{r|l} & \overline{46}\,\overline{24} \\ \hline & \end{array}$$

Step 2: For the first group, 46, find the perfect square just less than it. The closest perfect square that's less than 46 is 36. The square root of 36 is the first digit of your answer, so put down 6. To get the divisor, double 6: 6 × 2 = 12. Put the 12 on the left.

$$\begin{array}{r|l} 12 & \overline{46}\,\overline{24} \\ \hline & 6 \end{array}$$

Step 3: Subtract 46 minus the square of 6. The square of 6 is 36. So we have 46 – 36 = 10. This 10 is our remainder. We prefix 10 to 2 so it becomes 102.

$$\begin{array}{r|l} 12 & \overline{46}_{10}\overline{24} \\ \hline & 6 \end{array}$$

Step 4: Divide 102 by 12: 102 ÷ 12 = 8, remainder 6. Put down 8 and prefix the remainder to the 4 so it becomes 64.

$$\begin{array}{r|l} 12 & \overline{46}_{10}\overline{2_6 4} \\ \hline & 68 \end{array}$$

Step 5: Subtract 64 by the duplex of 8. In this case, the duplex of 8 is 8^2, or 64: $64 - 64 = 0$. Dividing 0 by 12 gives you 0, so you're finished. Because the answer only has two digits, the 0 goes after the decimal.

$$\begin{array}{r|l} 12 & \overline{46}_{10}\overline{2_{6}4} \\ \hline & 68.0 \end{array}$$

Solution: The answer is 68.0.

<u>**Example 3**</u>

Find $\sqrt{2656}$.

Step 1: Divide the number into groups of two from right to left. Because there are two groups, 26 and 56, the answer is going to contain two digits before the decimal.

$$\begin{array}{r|l} & \overline{26}\,\overline{56} \\ \hline & \end{array}$$

Step 2: For the first group, 26, find the perfect square just less than it. The closest perfect square that's less than 26 is 25. The square root of 25 is the first digit of your answer, so put down 5. To get the divisor, double 5: $5 \times 2 = 10$. Put the 10 on the left.

$$\begin{array}{r|l} 10 & \overline{26}\,\overline{56} \\ \hline & 5 \end{array}$$

Step 3: Subtract 26 minus the square of 5. The square of 5 is 25. So we have $26 - 25 = 1$. This 1 is our remainder which we prefix to 5, making it 15.

$$\begin{array}{r|l} 10 & \overline{26}_{1}\overline{56} \\ \hline & 5 \end{array}$$

Step 4: Divide 15 by 10: $15 \div 10 = 1$, remainder 5. Put down 1 and prefix the remainder to 6 so it becomes 56.

$$\begin{array}{r|l} 10 & \overline{26}_{1}\overline{5_{5}6} \\ \hline & 51 \end{array}$$

Step 5: Subtract 56 by the duplex of 1. In this case, the duplex of 1 is 1^2, or 1: 56 – 1 = 55. To get the number after the decimal, divide 55 by 10: 55 ÷ 10 = 5, remainder 5. Put down 5 after the decimal.

$$\begin{array}{r|l} 10 & \overline{26}{}_{1}\overline{5{}_{5}6} \\ \hline & 51.5 \end{array}$$

Solution: The answer is 51.5.

Cube Roots of Perfect Cubes

Perfect cubes are numbers whose cube roots are whole numbers. Like you did for perfect squares, you first need to check out the link between numbers 1 through 9 and their cubes.

Number	Cubes	Last Digit	Digit Sum
1	1	1	1
2	8	8	8
3	27	7	9
4	64	4	1
5	125	5	8
6	216	6	9
7	343	7	1
8	512	2	8
9	729	9	9

Here are the four connections you should notice:

1. The digit sums (see Chapter 5) go in a cycle of 1, 8, 9. The digit sum is 1 for 1, 4, and 7; 8 for 2, 5, and 8; and 9 for 3, 6, and 9.

2. If the cube ends in a 1, 4, 5, 6, or 9, its cube root always ends with that same number.

3. If the cube ends in a 8, the cube root ends in a 2; if the cube root ends in a 2, the cube ends in a 8.
4. If the cube ends in a 7, the cube root ends in a 3; if the cube ends in a 3, the cube root ends in a 7.

This information will help you find the cube roots.

Just because the number has a digit sum of 1, 8, or 9 doesn't necessarily mean it's a perfect cube; other cubes can have those digit sums. Keep that mind if you encounter numbers for which you have to find the cube root.

To find the cube roots of perfect cubes, you first split the number into groups of three to find out how many digits are in the answer. You then find the first part of the answer using the first group you made. The table can help you with the steps that involve finding the last digit of cube root based on the last digit of the cube.

<u>Example 1</u>

Find the cube root of a perfect cube 3,375.

Step 1: Split the number into groups of three from right to left to determine how many digits are in the answer. There are two groups, 3 and 375, so the answer is going to have two digits.

Step 2: Use the last digit of 3,375 to find the last digit of your answer. The last digit is 5, so according to the table, the last digit of the answer is 5.

Step 3: Look at the first group, 3, and figure out which cube value is less than or equal to it to get the first digit of your answer. The closest cube to 3 is 1. The cube root of 1 is 1, so 1 is the first digit.

Step 4: Compare the digit sums of the number and the answer to be sure they match.

$3375: 3 + 3 + 7 + 5 = 18; 1 + 8 = 9$

$15^3: (1 + 5)^3 = 6^3 = 216; 2 + 1 + 6 = 9$

Solution: The answer is 15.

Example 2

Find the cube root of the perfect cube 328,509.

Step 1: Split the number into groups of three from right to left to determine how many digits are in the answer. There are two groups, 328 and 509, so the answer is going to have two digits.

Step 2: Use the last digit of 328,509 to find the last digit of your answer. The last digit is 9, so according to the table, the last digit of the answer is 9.

Step 3: Look at the first group, 328, and figure out which cube value is less than or equal to it to get the first digit of your answer. As you can see via the table, the closest cube to 328 is 216. The cube root of 216 is 6, so 6 is the first digit.

Step 4: Compare the digit sums of the number and the answer to be sure they match.

$328509: 3 + 2 + 8 + 5 + 0 + 9 = 27; 2 + 7 = 9$

$69^3: (6 + 9)^3 = 15^3 = 3375; 3 + 3 + 7 + 5 = 18; 1 + 8 = 9$

Solution: The answer is 69.

Example 3

Find the cube root of the perfect cube 175,616.

Step 1: Split the number into groups of three from right to left to determine how many digits are in the answer. There are two groups, 175 and 616, so the answer is going to have two digits.

Step 2: Use the last digit of 175,616 to find the last digit of your answer. The last digit is 6, so according to the table, the last digit of the answer is 6.

Step 3: Look at the first group, 175, and figure out which cube value is less than or equal to it to get the first digit of your answer. As you can see via the table, the closest cube to 175 is 125. The cube root of 125 is 5, so 5 is the first digit.

Step 4: Compare of the number and the answer to be sure they match.

$175616: 1 + 7 + 5 + 6 + 1 + 6 = 26; 2 + 6 = 8$

$56^3: (5 + 6)^3 = 11^3 = 1331; 1 + 3 + 3 + 1 = 8$

Solution: The answer is 56.

Don't use this process for anything but perfect cubes, because you won't get the correct answer. Also, restrict this process to cubes of six digits or less; working larger numbers using this method can be an exhausting process!

The Least You Need to Know

- A perfect square should end in 1, 4, 5, 6, or 9.
- To find the square root of any number, divide by two times the first digit of your answer and subtract by the duplexes.
- If a perfect cube ends in 1, 4, 5, 6, or 9, the root should end with the same number. If the cube ends in 2, the root should end in 8, and vice versa. If the cube ends in 3, the root should end in 7, and vice versa.
- The digit sum can help you confirm that the answer you get matches the number.

APPENDIX

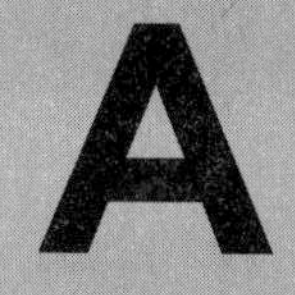

Practice Problems

Feel like practicing some of the methods in the book? This appendix provides some sample problems for you to try out using the different things you learned.

Chapter 1: Multiplication

Solve the following problems using finger multiplication.

1. 8×6
2. 7×7
3. 6×9
4. 10×9
5. 8×9

Solve the following problems using what you learned for multiplication by 11.

6. $2{,}610 \times 11$
7. $5{,}092 \times 11$
8. $3{,}475 \times 11$
9. $6{,}169 \times 11$
10. $9{,}251 \times 11$

Solve the following problems using 10 below the base method.

11. 9×8
12. 8×6
13. 9×7
14. 8×4
15. 7×8

Solve the following problems using 100 below the base method.

16. 95×98
17. 89×95
18. 92×87
19. 95×92
20. 98×88

Solve the following problems using 1,000 below the base method.

21. 997×990
22. 984×994
23. 988×987
24. 998×997
25. 994×993

Solve the following problems using 10 above the base method.

26. 14×19
27. 14×12
28. 19×17
29. 12×19
30. 11×13

Solve the following problems using 100 above the base method.

31. 119×103
32. 116×106
33. 118×116
34. 112×113
35. 110×113

Solve the following problems using 1,000 above the base method.

36. $1{,}007 \times 1{,}009$
37. $1{,}016 \times 1{,}003$
38. $1{,}015 \times 1{,}018$
39. $1{,}002 \times 1{,}012$
40. $1{,}018 \times 1{,}002$

Solve the following problems using the 10 above and below the base method.

41. 11×9

42. 15×8

43. 12×8

44. 13×7

45. 18×6

Solve the following problems using the 100 above and below the base method.

46. 101×98

47. 105×88

48. 111×97

49. 108×94

50. 112×96

Solve the following problems using the 1,000 above and below the base method.

51. $1{,}001 \times 996$

52. $1{,}008 \times 988$

53. $1{,}011 \times 967$

54. $1{,}013 \times 999$

55. $1{,}051 \times 998$

Solve the following problems using the method you learned for multiples and submultiples.

56. 54×56

57. 215×204

58. 299×296

59. 301×305

60. 67×65

Solve the following problems using the method you learned for combined problems.

61. 211×198

62. 56×47

63. 107×94

64. 501×489

65. 317×283

Solve the following problems using the vertically and crosswise method for two-digit numbers.

66. $\begin{array}{r} 97 \\ \times\ 44 \\ \hline \end{array}$

67. $\begin{array}{r} 80 \\ \times\ 18 \\ \hline \end{array}$

68. $\begin{array}{r} 43 \\ \times\ 86 \\ \hline \end{array}$

69. $\begin{array}{r} 27 \\ \times\ 52 \\ \hline \end{array}$

70. $\begin{array}{r} 54 \\ \times\ 11 \\ \hline \end{array}$

Solve the following problems using the vertically and crosswise method for three-digit numbers.

71. $\begin{array}{r} 397 \\ \times\ 993 \\ \hline \end{array}$

72. $\begin{array}{r} 904 \\ \times\ 149 \\ \hline \end{array}$

73. $\begin{array}{r} 980 \\ \times\ 780 \\ \hline \end{array}$

74. $\begin{array}{r} 394 \\ \times\ 531 \\ \hline \end{array}$

75. $\begin{array}{r} 766 \\ \times\ 626 \\ \hline \end{array}$

Chapter 2: Addition

Solve the following problems using left-to-right addition for two-digit numbers.

1. $\begin{array}{r} 59 \\ +\ 91 \\ \hline \end{array}$

2. $\begin{array}{r} 62 \\ +\ 95 \\ \hline \end{array}$

3. $\begin{array}{r} 71 \\ +\ 99 \\ \hline \end{array}$

4. $\begin{array}{r} 59 \\ +\ 89 \\ \hline \end{array}$

5. $\begin{array}{r} 52 \\ +\ 84 \\ \hline \end{array}$

Solve the following problems using left-to-right addition for three-digit numbers.

6. $\begin{array}{r} 377 \\ +\ 787 \\ \hline \end{array}$

7. $\begin{array}{r} 256 \\ +\ 653 \\ \hline \end{array}$

8. $\begin{array}{r} 945 \\ +\ 114 \\ \hline \end{array}$

9. $\begin{array}{r} 336 \\ +\ 870 \\ \hline \end{array}$

10. $\begin{array}{r} 610 \\ +\ 759 \\ \hline \end{array}$

Solve the following problems using left-to-right columnar addition.

11. $\begin{array}{r} 4{,}275 \\ 7{,}883 \\ 5{,}848 \\ +\ 2{,}784 \\ \hline \end{array}$

12. $\begin{array}{r} 4{,}713 \\ 6{,}046 \\ 4{,}647 \\ +\ 4{,}104 \\ \hline \end{array}$

13. $\begin{array}{r} 5{,}448 \\ 8{,}520 \\ 6{,}446 \\ +\ 1{,}217 \\ \hline \end{array}$

14. $\begin{array}{r} 2{,}979 \\ 7{,}546 \\ 4{,}577 \\ +\ 4{,}623 \\ \hline \end{array}$

15. $\begin{array}{r} 2{,}467 \\ 2{,}629 \\ 1{,}159 \\ +\ 5{,}520 \\ \hline \end{array}$

Solve the following problems using the addition method for numbers near 10 or a multiple of 10.

16. $57 + 6$

17. $64 + 6$

18. $68 + 8$

19. $66 + 8$

20. $77 + 7$

Solve the following problems using number splitting.

21. $6{,}705 + 1{,}856$

22. $7{,}369 + 6{,}635$

23. $9{,}588 + 2{,}233$

24. $7{,}403 + 1{,}790$

25. $9{,}915 + 4{,}281$

Chapter 3: Subtraction

Solve the following problems using left-to-right subtraction.

1. $410 - 127$

2. $479 - 444$

3. $651 - 471$

4. $613 - 174$

5. $386 - 101$

Solve the following problems using the method for numbers near 10 or a multiple of 10.

6. $44 - 7$

7. $21 - 7$

8. $50 - 9$

9. $22 - 8$

10. $49 - 7$

Solve the following problems using number splitting.

11. $6{,}862 - 2{,}728$

12. $7{,}995 - 1{,}293$

13. $8{,}226 - 5{,}451$

14. $5{,}293 - 3{,}217$

15. $6{,}154 - 4{,}696$

Solve the following problems using the all from 9 and last from 10 method.

16. $\begin{array}{r} 57{,}863 \\ -\ 48{,}880 \\ \hline \end{array}$

17. $\begin{array}{r} 25{,}688 \\ -\ 21{,}375 \\ \hline \end{array}$

18. $\begin{array}{r} 92{,}198 \\ -\ 25{,}345 \\ \hline \end{array}$

19. $\begin{array}{r} 81{,}825 \\ -\ 28{,}447 \\ \hline \end{array}$

20. $\begin{array}{r} 94{,}114 \\ -\ 23{,}171 \\ \hline \end{array}$

Chapter 4: Division

Solve the following problems using the flag method for two-digit divisors.

1. $4{,}305 \div 80 =$ ____
2. $5{,}067 \div 25 =$ ____
3. $6{,}033 \div 28 =$ ____
4. $4{,}649 \div 93 =$ ____
5. $5{,}657 \div 73 =$ ____

Solve the following problems using the flag method for three-digit divisors.

6. $78{,}115 \div 767 =$ ____
7. $32{,}316 \div 143 =$ ____
8. $66{,}892 \div 141 =$ ____
9. $16{,}771 \div 278 =$ ____
10. $69{,}027 \div 842 =$ ____

Solve the following problems using the flag method for four-digit divisors.

11. $796{,}267 \div 8{,}667 =$ ____
12. $717{,}219 \div 7{,}682 =$ ____
13. $898{,}376 \div 3{,}442 =$ ____
14. $881{,}306 \div 7{,}256 =$ ____
15. $653{,}909 \div 3{,}595 =$ ____

Solve the following problems using auxiliary fractions.

16. $\frac{1}{19}$

17. $\frac{1}{17}$

18. $\frac{12}{29}$

19. $\frac{1}{49}$

20. $\frac{1}{59}$

Chapter 5: Checking Your Answers with Digit Sums

See if the following problems are true or false using digit sums.

1. $3{,}756 + 3{,}728 = 7{,}484$
2. $746 + 129 = 874$
3. $382 + 459 = 841$
4. $2{,}935 + 2{,}845 = 5{,}779$
5. $1{,}948 + 6{,}849 = 8{,}797$
6. $432 - 287 = 143$
7. $567 - 398 = 169$
8. $987 - 234 = 750$
9. $1{,}578 - 1{,}266 = 312$
10. $2{,}399 - 1{,}166 = 1{,}232$
11. $123 \times 340 = 41{,}820$
12. $214 \times 322 = 68{,}901$
13. $111 \times 144 = 15{,}984$
14. $433 \times 221 = 95{,}692$
15. $321 \times 554 = 177{,}834$
16. $420 \div 21 = 20$
17. $375 \div 75 = 5$
18. $243 \div 3 = 82$
19. $343 \div 7 = 49$
20. $570 \div 19 = 30$

Chapter 6: Fractions

Solve the following problems using what you learned for adding fractions.

1. $\frac{5}{7} + \frac{3}{8} =$ ____
2. $\frac{7}{11} + \frac{4}{9} =$ ____
3. $\frac{11}{13} + \frac{6}{11} =$ ____
4. $\frac{1}{6} + \frac{3}{16} =$ ____
5. $\frac{17}{31} + \frac{1}{8} =$ ____

Solve the following problems using what you learned for subtracting fractions.

6. $\frac{1}{5} - \frac{1}{7} =$ ___

7. $\frac{11}{15} - \frac{4}{7} =$ ___

8. $\frac{9}{20} - \frac{2}{5} =$ ___

9. $\frac{1}{19} - \frac{6}{199} =$ ___

10. $\frac{7}{12} - \frac{11}{20} =$ ___

Solve the following problems using what you learned for multiplying fractions.

11. $\frac{7}{9} \times \frac{3}{7} =$ ___

12. $\frac{1}{4} \times \frac{9}{11} =$ ___

13. $\frac{3}{13} \times \frac{7}{19} =$ ___

14. $\frac{2}{15} \times \frac{25}{60} =$ ___

15. $\frac{9}{14} \times \frac{11}{12} =$ ___

Solve the following problems using what you learned for dividing fractions.

16. $\frac{1}{7} \div \frac{5}{14} =$ ___

17. $\frac{9}{10} \div \frac{12}{18} =$ ___

18. $\frac{1}{14} \div \frac{5}{7} =$ ___

19. $\frac{34}{17} \div \frac{17}{34} =$ ___

20. $\frac{50}{18} \div \frac{10}{30} =$ ___

Chapter 7: Decimals

Solve the following problems using what you learned for adding decimals.

1. $5.37 + 9.14$
2. $0.85 + 9.83$
3. $0.58 + 2.45$
4. $0.45 + 8.23$
5. $0.35 + 9.41$

Solve the following problems using what you learned for subtracting decimals.

6. $91.43 - 85.78$
7. $35.58 - 23.79$
8. $93.98 - 87.78$
9. $93.98 - 87.78$
10. $79.32 - 19.42$

Solve the following problems using what you learned for multiplying decimals.

11. 10.0×6.8
12. 3.0×9.9
13. 1.6×2.6
14. 1.6×5.2
15. 4.7×9.9

Solve the following problems using what you learned for dividing decimals.

16. $9.8 \div 0.7 =$ ____
17. $1.69 \div 1.3 =$ ____
18. $4.84 \div 0.22 =$ ____
19. $34.3 \div 0.07 =$ ____
20. $72.9 \div 0.081 =$ ____

Chapter 8: Percentages

Convert the following percentages into fractions.

1. 34%
2. 49%
3. 97%
4. 86%
5. 18%

Convert the following fractions into percentages.

6. $\frac{3}{19}$ 7. $\frac{5}{25}$ 8. $\frac{10}{40}$

9. $\frac{5}{17}$ 10. $\frac{3}{19}$

Convert the following percentages to decimals.

11. 24.5% 12. 12.4% 13. 35%

14. 96% 15. 26.88%

Find the percentage of the given quantity.

16. 23% of 500 17. 14% of 345

18. 36.5% of 900 19. 30% of 458

20. 56.7% of 100

Express one quantity as a percentage of the other in the following problems.

21. Express 24 as a percentage of 48.
22. Express 38 as a percentage of 228.
23. Express 52 as a percentage of 235.
24. Express 20 as a percentage of 465.
25. Express 34 as a percentage of 436.

Approximate the percentages for the following problems.

26. Estimate 54 as a percentage of 234.
27. Estimate 16 as a percentage of 256.
28. Estimate 56 as a percentage of 456.
29. Estimate 34 as a percentage of 250.
30. Estimate 20 as a percentage of 458.

Chapter 9: Divisibility

Solve the following problems using the osculation method.

1. Find out if 7,680 is divisible by 80.
2. Find out if 1,798 is divisible by 62.
3. Find out if 1,785 is divisible by 17.
4. Find out if 4,368 is divisible by 91.
5. Find out if 3,024 is divisible by 54.
6. Find out if 2,756 is divisible by 52.
7. Find out if 5,487 is divisible by 59.
8. Find out if 8,905 is divisible by 65.
9. Find out if 4,617 is divisible by 81.
10. Find out if 3,198 is divisible by 13.
11. Find out if 3,055 is divisible by 65.
12. Find out if 6,540 is divisible by 15.
13. Find out if 2,166 is divisible by 38.
14. Find out if 4,386 is divisible by 43.
15. Find out if 2,880 is divisible by 60.

Chapter 10: Squared Numbers

Solve the following problems using what you learned for finding squares of numbers ending in 5.

1. $75^2 =$ ____ 2. $25^2 =$ ____ 3. $55^2 =$ ____

4. $45^2 =$ ____ 5. $65^2 =$ ____

Solve the following problems using what you learned for finding squares of numbers ending in 50.

6. $52^2 =$ ____ 7. $55^2 =$ ____ 8. $53^2 =$ ____

9. $46^2 =$ ____ 10. $45^2 =$ ____

Solve the following problems using the duplex method for two-digit squares.

11. $82^2 =$ ____ 12. $26^2 =$ ____ 13. $12^2 =$ ____

14. $84^2 =$ ____ 15. $78^2 =$ ____

Solve the following problems using the duplex method for three-digit squares.

16. $651^2 =$ ____ 17. $937^2 =$ ____ 18. $990^2 =$ ____

19. $363^2 =$ ____ 20. $366^2 =$ ____

Solve the following problems using the duplex method for four-digit squares.

21. $1{,}234^2$ 22. $5{,}482^2$ 23. $7{,}011^2$

24. $6{,}321^2$ 25. $1{,}082^2$

Solve the following problems using what you learned for combined operations.

26. $64^2 + 92^2 =$ ____ 27. $43^2 + 71^2 =$ ____

28. $29^2 + 91^2 =$ ____ 29. $88^2 + 67^2 =$ ____

30. $77^2 + 91^2 =$ ____

Chapter 11: Cubed Numbers

Solve the following problems using what you learned for finding the cubes of two-digit numbers.

1. $54^3 =$ ____ 2. $31^3 =$ ____

3. $69^3 =$ ____ 4. $30^3 =$ ____

5. $45^3 =$ ____

Solve the following problems using what you learned for calculating cubes near a base.

6. $111^3 =$ ____ 7. $100^3 =$ ____

8. $109^3 =$ ____ 9. $114^3 =$ ____

10. $102^3 =$ ____

Chapter 12: Numbers to the Fourth and Fifth Powers

Solve the following problems using what you learned for calculating numbers to the fourth power.

1. $79^4 =$ ____
2. $34^4 =$ ____
3. $70^4 =$ ____
4. $96^4 =$ ____
5. $23^4 =$ ____

Solve the following problems using what you learned for calculating numbers to the fifth power.

6. $28^5 =$ ____
7. $13^5 =$ ____
8. $42^5 =$ ____
9. $36^5 =$ ____
10. $18^5 =$ ____

Chapter 13: Square and Cube Roots

Solve the following problems using what you learned for finding the square roots of perfect squares.

1. $\sqrt{3249} =$ ____
2. $\sqrt{7056} =$ ____
3. $\sqrt{8464} =$ ____
4. $\sqrt{5476} =$ ____
5. $\sqrt{3969} =$ ____

Solve the following problems using the duplexes to find square roots.

6. $\sqrt{9604} =$ ____
7. $\sqrt{7744} =$ ____
8. $\sqrt{2916} =$ ____
9. $\sqrt{6561} =$ ____
10. $\sqrt{4761} =$ ____

Solve the following problems using what you learned for finding the cube roots of perfect cubes.

11. $\sqrt[3]{68921} =$ ____
12. $\sqrt[3]{59319} =$ ____
13. $\sqrt[3]{166375} =$ ____
14. $\sqrt[3]{551368} =$ ____
15. $\sqrt[3]{10648} =$ ____

APPENDIX

B

Answer Key

This appendix includes the answers to the practice problems in Appendix A. Check your work and see how you've progressed!

Chapter 1: Multiplication

1.	48	2.	49	3.	54
4.	90	5.	42	6.	28,710
7.	56,012	8.	38,225	9.	7,859
10.	101,761	11.	72	12.	48
13.	63	14.	32	15.	56
16.	9,310	17.	8,455	18.	8,004
19.	8,740	20.	8,624	21.	987,030
22.	978,096	23.	975,156	24.	995,006
25.	987,042	26.	266	27.	168
28.	323	29.	228	30.	143
31.	12,257	32.	12,296	33.	13,688
34.	12,656	35.	12,430	36.	1,016,063

37.	1,019,048	38.	1,033,270	39.	1,014,024
40.	1,020,036	41.	99	42.	120
43.	96	44.	91	45.	108
46.	9,898	47.	9,240	48.	10,767
49.	10,152	50.	10,752	51.	996,996
52.	995,904	53.	977,637	54.	1,011,987
55.	1,048,898	56.	3,024	57.	43,860
58.	88,504	59.	91,805	60.	4,355
61.	41,778	62.	2,632	63.	10,058
64.	244,989	65.	89,711	66.	4,268
67.	1,120	68.	3,698	69.	1,405
70.	594	71.	394,221	72.	134,696
73.	764,400	74.	209,214	75.	479,516

Chapter 2: Addition

1.	150	2.	157	3.	170
4.	148	5.	136	6.	1,164
7.	909	8.	1,059	9.	1,206
10.	1,369	11.	20,790	12.	19,510
13.	21,631	14.	19,725	15.	12,135
16.	63	17.	70	18.	76
19.	74	20.	84	21.	8,561
22.	14,004	23.	11,821	24.	9,193
25.	14,196				

Chapter 3: Subtraction

1.	283	2.	35	3.	180
4.	439	5.	285	6.	37
7.	14	8.	41	9.	14
10.	42	11.	4,134	12.	6,702
13.	2,775	14.	2,076	15.	1,458
16.	8,983	17.	4,313	18.	66,853
19.	53,378	20.	70,943		

Chapter 4: Division

1.	53.813	2.	202.68	3.	215.464
4.	49.989	5.	77.493	6.	101.845
7.	225.986	8.	474.411	9.	60.327
10.	81.980	11.	91.873	12.	93.364
13.	261.004	14.	121.459	15.	181.894
16.	0.0526	17.	0.0588	18.	0.4137
19.	0.020408	20.	0.01694		

Chapter 5: Checking Your Answers with Digit Sums

1.	True	2.	False	3.	True
4.	False	5.	True	6.	False
7.	True	8.	False	9.	True
10.	False	11.	True	12.	False
13.	True	14.	False	15.	True
16.	True	17.	True	18.	False
19.	True	20.	True		

Chapter 6: Fractions

1.	$\frac{61}{56}$	2.	$\frac{107}{99}$	3.	$\frac{199}{143}$
4.	$\frac{17}{48}$	5.	$\frac{167}{248}$	6.	$\frac{2}{35}$
7.	$\frac{17}{105}$	8.	$\frac{1}{20}$	9.	$\frac{85}{3781}$
10.	$\frac{1}{30}$	11.	$\frac{1}{3}$	12.	$\frac{9}{44}$
13.	$\frac{21}{247}$	14.	$\frac{1}{18}$	15.	$\frac{99}{168}$
16.	$\frac{2}{5}$	17.	$\frac{27}{20}$	18.	$\frac{1}{10}$
19.	4	20.	$\frac{25}{3}$		

Chapter 7: Decimals

1.	14.51	2.	10.68	3.	3.03
4.	8.68	5.	9.76	6.	5.65
7.	11.79	8.	6.20	9.	21.26
10.	59.90	11.	68.00	12.	29.70
13.	4.16	14.	8.32	15.	46.53
16.	21.035	17.	3.312	18.	7.494
19.	5.792	20.	16.117		

Chapter 8: Percentages

1.	$\frac{17}{50}$	2.	$\frac{49}{100}$	3.	$\frac{97}{100}$
4.	$\frac{43}{50}$	5.	$\frac{9}{50}$	6.	50%
7.	40%	8.	52%	9.	65%
10.	27%	11.	0.245	12.	0.124
13.	0.35	14.	0.96	15.	0.2688
16.	150	17.	48.3	18.	328.5
19.	137.4	20.	56.7	21.	50%
22.	16.67%	23.	22.13%	24.	47%
25.	7.79%	26.	23.07%	27.	6.25%
28.	12.28%	29.	13.6%	30.	4.36%

Chapter 9: Divisibility

1. Yes, 7,680 is divisible by 80.
2. Yes, 1,798 is divisible by 62.
3. Yes, 1,785 is divisible by 17.
4. Yes, 4,368 is divisible by 91.
5. Yes, 3,024 is divisible by 54.
6. Yes, 2,756 is divisible by 52.
7. Yes, 5,487 is divisible by 59.
8. Yes, 8,905 is divisible by 65.
9. Yes, 4,617 is divisible by 81.
10. Yes, 3,198 is divisible by 13.
11. Yes, 3,055 is divisible by 65.
12. Yes, 6,540 is divisible by 15.
13. Yes, 2,166 is divisible by 38.
14. Yes, 4,386 is divisible by 43.
15. Yes, 2,880 is divisible by 60.

Chapter 10: Squared Numbers

1. 5,625
2. 625
3. 3,025
4. 2,025
5. 4,225
6. 2,704
7. 3,025
8. 2,809
9. 2,116
10. 2,025
11. 6,724
12. 676
13. 144
14. 7,056
15. 6,084
16. 423,801
17. 877,969
18. 980,100
19. 131,769
20. 133,956
21. 1,522,756
22. 30,052,324
23. 49,154,121
24. 39,955,041
25. 1,170,724
26. 12,560
27. 6,890
28. 9,122
29. 12,233
30. 14,210

Chapter 11: Cubed Numbers

1. 157,464
2. 29,791
3. 328,509
4. 27,000
5. 91,125
6. 1,367,631
7. 1,000,000
8. 1,295,029
9. 1,481,544
10. 1,061,208

Chapter 12: Numbers to the Fourth and Fifth Powers

1. 38,950,081
2. 1,336,336
3. 24,010,000
4. 84,934,656
5. 279,841
6. 17,210,368
7. 371,293
8. 130,691,232
9. 60,466,176
10. 1,889,568

Chapter 13: Square and Cube Roots

1.	28	2.	16	3.	14
4.	29	5.	27	6.	261
7.	582	8.	306	9.	252
10.	212	11.	27	12.	36
13.	95	14.	99	15.	46

Index

A

B

C

D-E

F-H

I–K

L

M

N

O

P

Q

R

S

T

U

V-Z